ANATOMY OF THE

HUMAN EYE

A COLORING ATLAS

Lisa A Ostrin

1ST EDITION

Anatomy of the Human Eye: A Coloring Atlas
First Edition
ISBN 9798535235509

Author and Illustrator

Lisa A Ostrin, OD, PhD, FAAO, FARVO
Associate Professor
University of Houston College of Optometry

Special thanks to colleagues who reviewed and edited sections of this book

Jan PG Bergmanson, OD, PhD, DSc, FAAO
Professor
University of Houston College of Optometry

Han Cheng, OD, PhD
Clinical Professor
University of Houston College of Optometry

Thomas Frank Freddo, OD, PhD, FAAO, FARVO
Professor
School of Optometry
Massachusetts College of Pharmacy and Health Sciences

Laura Frishman, PhD, FAAO, FARVO
Professor
Associate Dean for Graduate Studies and Research
University of Houston College of Optometry

DEDICATION

This book is dedicated to my family. To Edwin for inspiring me in science, medicine, and anatomy every day, for always believing in me, and for always humoring me. To my children for their support, understanding, and helpful suggestions in this process. In addition, thanks to Edwin for modeling the facial nerve, Ronnie for modeling the external eye, Ben for modeling the facial muscles, Lila Bean for modeling the skull, Noe for modeling the paranasal sinuses, and Ashby for modeling the myelin sheath. To my parents, grandparents, and brother for providing a legacy of science, medicine, and art, and to my in-laws for unwavering support.

INTRODUCTION

One of the most effective ways to learn anatomy is to draw anatomy. This book was designed to systematically cover all of the structures of the human eye. The eye is one of the most complex organs of the body. In this book, each structure is broken down into anatomically correct templates, so that beginning learners, students, educators, practitioners, and researchers in health professions, including optometry, medical, dental, and veterinary medicine, can benefit from the material presented in this atlas. In working through this book, I hope that you not only learn anatomy, but also enjoy creating annotated images that equally serve as a study guide and as art.

The figures and text in this book step through the eye in a logical and comprehensive manner. References for the pages in this book include figures from classic anatomists such as Hogan, Alvarado, and Weddell and Wolff, as well as my dissections and histological preparations. To use this book, I suggest filling in the labels and reading the corresponding descriptions about each image. Color and shade as you see fit. You may want to be consistent across images with your colors – red for arteries, blue for veins, yellow for nerves, pink for muscle – or, take creative freedom and express yourself!

About the author: Lisa Ostrin is an Associate Professor at the University of Houston College of Optometry. Dr. Ostrin received a Bachelor of Arts in Fine Art at the University of Texas at Austin and a Doctorate of Optometry and PhD in physiological optics at the University of Houston. From there, she completed post-doctoral work at Johns Hopkins University with a focus on retinal prosthetics and held a clinical researcher position at the University of California Berkeley School of Optometry investigating myopia and eye growth. She then returned to the University of Houston as a full-time faculty member. Dr. Ostrin gained further training by participating in the Anatomy Training Programme through the Anatomical Society of the UK. Dr. Ostrin teaches courses in the professional optometry and graduate studies programs, including topics such as human gross anatomy, ocular anatomy, accommodative physiology and pharmacology, and emmetropization and myopia. Her research focuses on understanding environmental influences on circadian rhythm and eye growth and developing treatments for myopia. She is a fellow of the American Academy of Optometry and a gold fellow of the Association for Research in Vision and Ophthalmology.

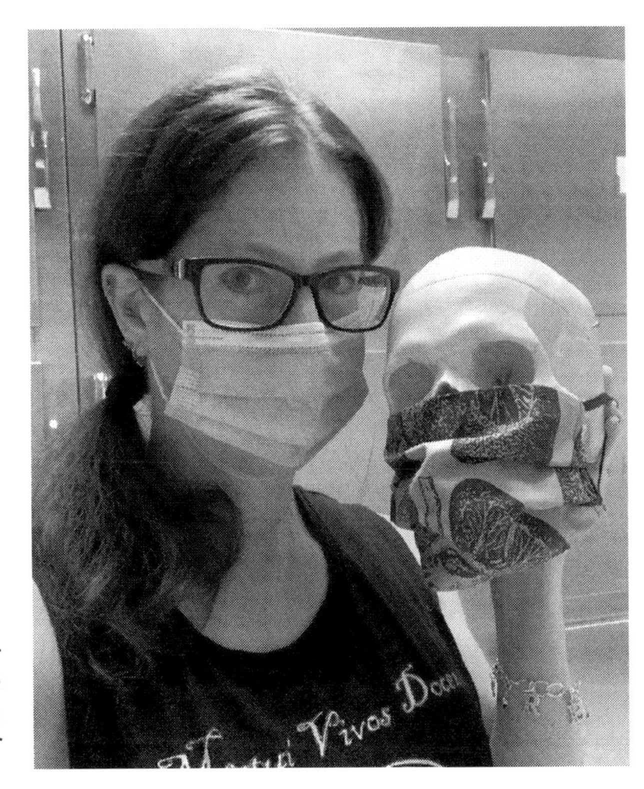

While moving from graduate school through post-doctoral positions and back to Houston as faculty, she and husband, a physician researcher, had four children (featured in these drawings). She is an advocate for work-life balance and integration for young scientists.

Table of Contents

Overview of the Eye and Adnexa

Overview of the Eye

The eye is a highly specialized organ that allows us to see the world around us. The eye consists of unique tissues with millions of cells that, together, refract light and convert photons into a signal that is ultimately sent to the visual cortex and perceived as an image.

The adult eye is approximately spherical and about 24 mm in length with a volume of 5-6 mL. The eye has three layers, or tunics, including the corneoscleral, uveal, and neurosensory layers. The outermost corneoscleral layer consists of the cornea and sclera. These are fibrous tissues which serve to provide shape to the eye and protect deeper structures. The middle uveal layer is nutritive. It is highly vascularized and highly pigmented, consisting of the iris, ciliary body, and choroid. The innermost neurosensory layer functions to detect light. This layer includes the retinal pigment epithelium and retina.

The eye is divided into two segments, anterior and posterior. The anterior segment spans from the cornea to posterior lens, and the posterior segment spans from the posterior lens to the back of the eye. Furthermore, the anterior segment is divided into two chambers; the anterior chamber spans from the posterior cornea to iris, and the posterior chamber spans from the iris to posterior lens. The anterior chamber is approximately 3.6 mm in depth centrally, and contains approximately 170-200 µL of fluid. The posterior chamber contains the crystalline lens, and has a volume of approximately 60 µL.

As light passes from the front of the eye to the retina, it is first refracted by the cornea, passes through the anterior chamber and pupil of the iris, is further refracted by the lens, then passes through the vitreous and inner retina to be captured by the photoreceptors. The signal will leave the eye via the optic nerve, where the majority of axons will synapse in the lateral geniculate nucleus of the midbrain, then the signal will be carried to the primary visual cortex via the optic radiations.

Figure Description

Transverse cross section of eye

Key

A anterior chamber
B posterior chamber
C anterior segment
D posterior segment
E cornea
F conjunctiva
G pupil
H iris
I crystalline lens
J ciliary body
K ora serrata
L sclera
M choroid
N retina
O optic nerve
P central retinal vasculature

Overview of the Eye

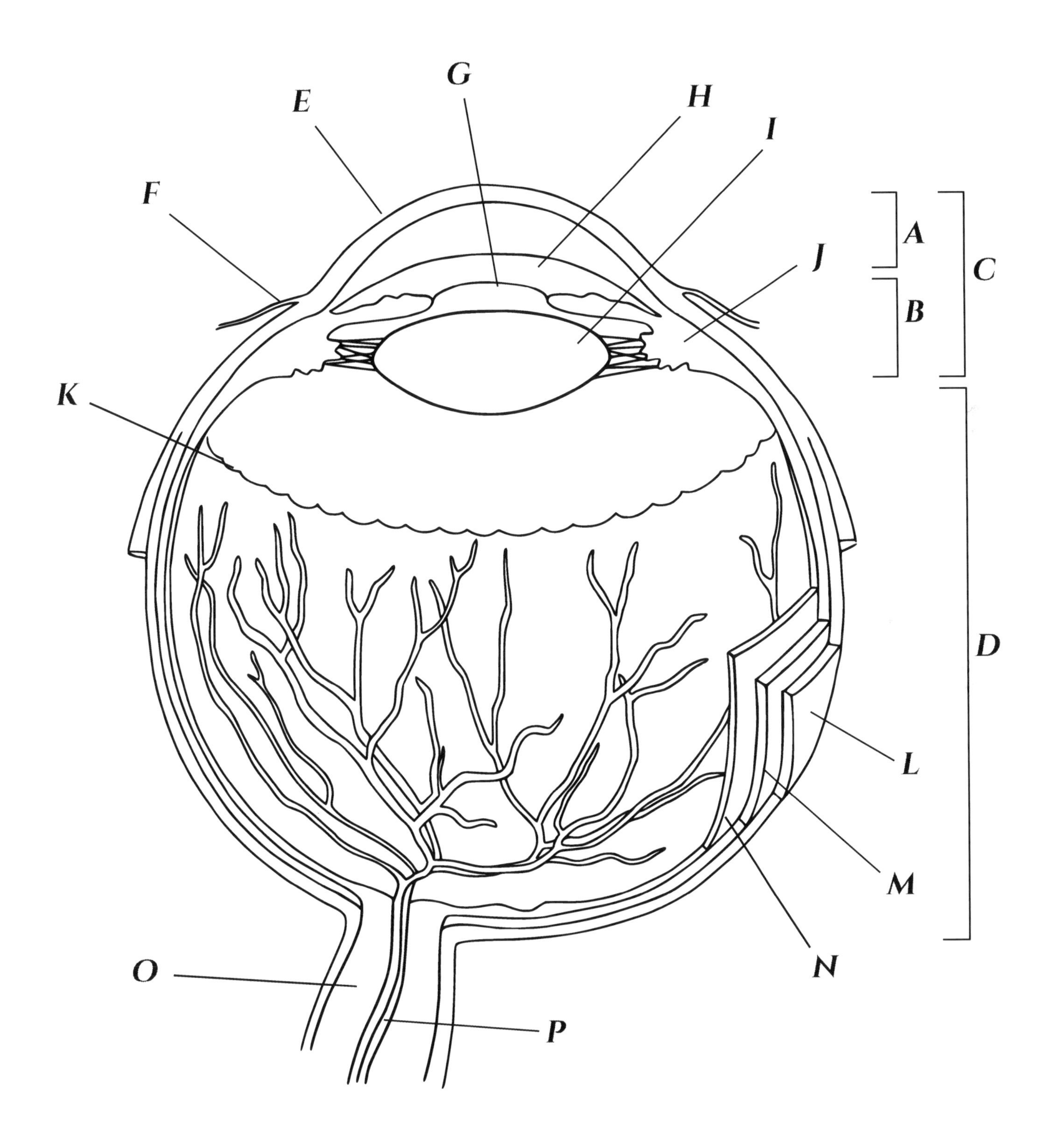

External Landmarks

The clinical exam begins with observing the external landmarks surrounding the eye. The eyelids span from the eyebrow superiorly to the malar and nasojugal sulci inferiorly. The eyebrow is important for protection of the eye and functions in facial expression. The sulci, or fissures, are where the skin is adhered to the underlying periosteum of the bone.

The eyelids are divided into orbital and palpebral regions. The orbital region is the outer portion, and the palpebral region is over the eye. The orbital and palpebral regions are divided by the superior and inferior orbital fissures. The superior orbital fissure is formed by the insertion of the aponeurosis of the levator palpebrae superioris muscle into the skin.

The superior and inferior eyelids meet at the medial and lateral canthi. The elliptical opening between the eyelids is the palpebral fissure, which spans approximately 30 mm horizontally and 10 mm vertically in an adult open eye. At the medial canthus, the caruncle and plica semilunaris are found. The caruncle is a mound of modified skin tissue that contains sebaceous and sweat glands, and may contain hair follicles. The plica semilunaris is a flat fold of bulbar conjunctiva that is a remnant of a third eyelid, or nictitating membrane. Some animals, including some fish and birds, still maintain the nictitating membrane.

The lid margins are the free edges of the eyelids and are approximately 2 mm in thickness. The anterior edge of the lid margin has 2-3 rows of cilia, or eyelashes, with approximately 150 eyelashes on the superior lid margin and 75 eyelashes on the inferior lid margin. Each eyelash takes about 10 weeks to grow and lasts about 5 months.

The lid margins contain numerous gland orifices that contribute to the tear film. The glands of Zeis and Moll are associated with the cilia follicles and the tarsal gland (aka Meibomian gland) orifices are found more posterior along the lid margin.

Figure Description

Anterior view of right eye

Key

A eyebrow
B upper eyelid, orbital portion
C upper eyelid, palpebral portion
D cilia (eyelashes)
E palpebral fissure
F lower eyelid palpebral portion
G lower eyelid, orbital portion
H lateral canthus
I medial canthus
J caruncle
K plica semilunaris
L superior orbital fissure
M inferior orbital fissure
N nasojugal sulcus
O malar sulcus

External Landmarks

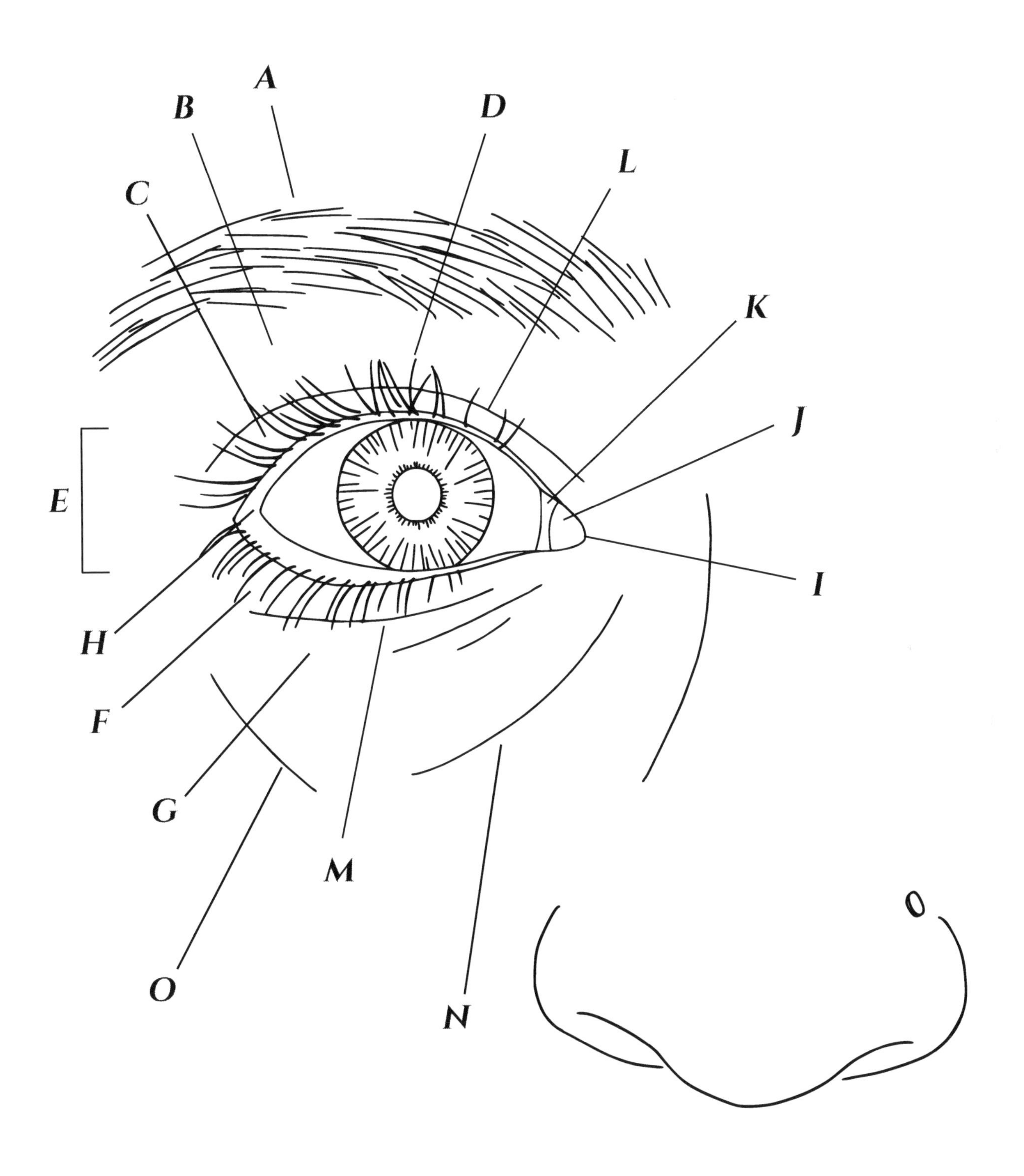

Muscles of Facial Expression

The muscles of facial expression are flat skeletal muscles that are found just deep to the skin of the face. The muscles of facial expression include several muscles around the eye and of the eyelid. These muscles insert into the skin and produce facial expressions and eyelid closure. They are innervated by the motor component of the facial nerve (CN VII). Blood supply to the facial muscles is via the facial artery, a branch of the external carotid artery.

The facial muscles are categorized by location and function, and include the buccolabial group around the mouth, nasal group around the nose, epicranial group around the cranium and neck, auricular group around the external ear, and orbital group around the eye.

The frontal belly of the occipitofrontalis muscle, also known as the frontalis muscle, spans across the forehead and is responsible for raising the eyebrows. The muscle fibers run vertically and span from the scalp to the superior orbital rim. The procerus muscle is found medially and is responsible for the transverse furrow between the eyes. The corregator supracilii muscles arch along the superior orbital rim, deep to the eyebrows, and serve to bring the eyebrows together.

The large circular muscle of the eyelids is the orbicularis oculi. This muscle is responsible for closing the eyelids, and can be divided into three regions - the orbital region is the outer portion, the palpebral or tarsal region is over the eye, and the ciliary region, also known as the muscle of Riolan, is along the lid margin. The muscle of Riolin serves to hold the lid margin taught to the eye. The medial portion of the palpebral region surrounding the lacrimal drainage apparatus is Horner's muscle, which aids in tear drainage from the ocular surface by pulling the canaliculi during blinking.

Figure Description

Anterior face, skin removed revealing superficial muscles of facial expression

Key

A galea aponeurosis
B frontal belly of occipitofrontalis muscle
C procerus muscle
D corrugator supracilii muscle
E orbicularis oculi muscle
F nasalis muscle
G alaque nasi muscle
H levator labii superioris muscle
I zygomaticus minor muscle
J zygomaticus major muscle
K levator anguli oris muscle
L buccinator muscle
M risorius muscle
N depressor anguli oris muscle
O platysma muscle
P depressor labii inferioris muscle
Q mentalis muscle

Facial Muscles

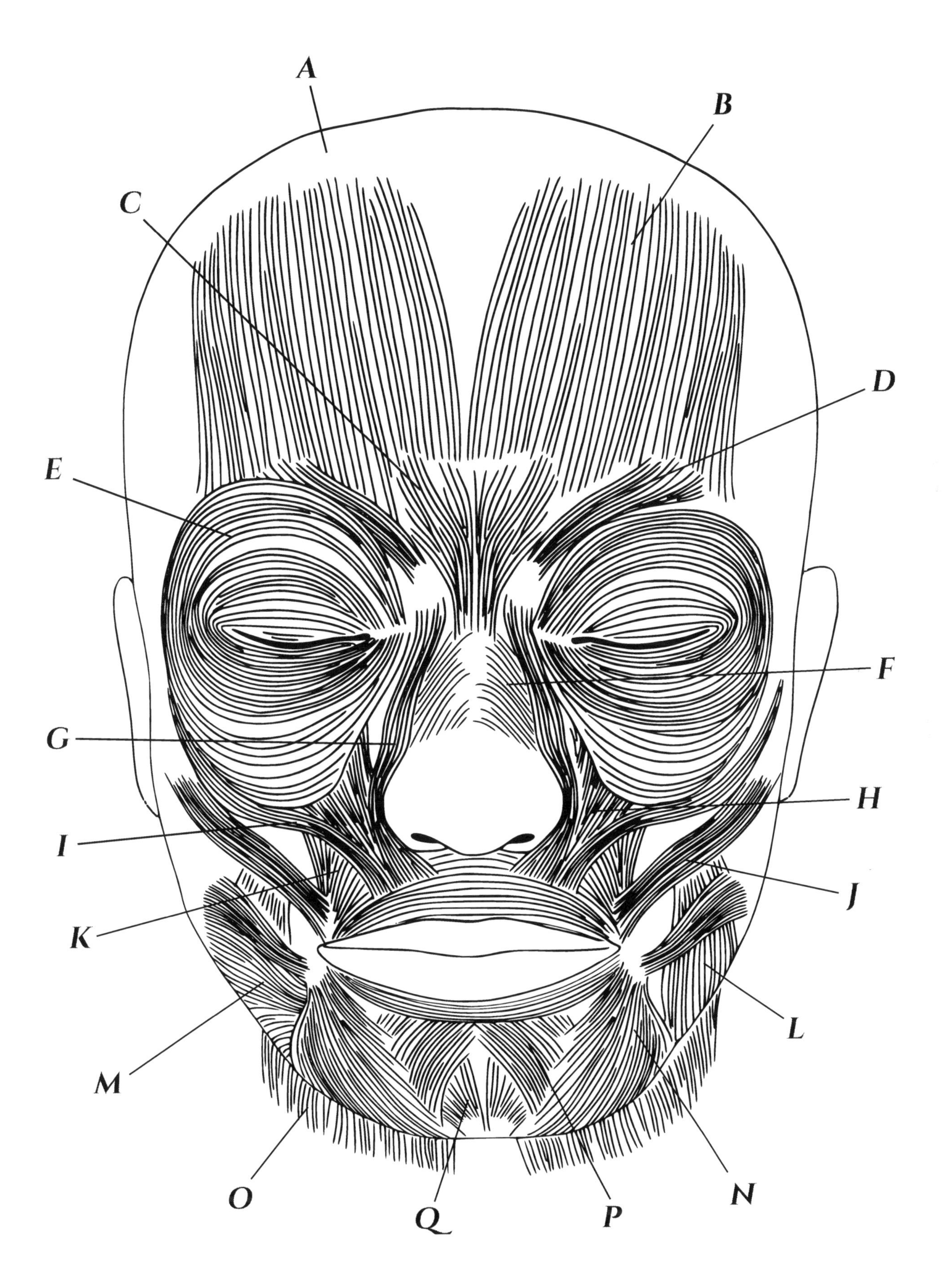

The Skull

The human skull consists of 22 bones that protect the brain and other sensory structures and provide structure to the face and jaw. All of the bones are fused together except for the mandible. The skull contains numerous foramina, or passageways, for nerves and vasculature, the largest being the foramen magnum, through which the spinal cord passes. Several fossae, which are depressions or cavities, provide protection for various structures. The skull also contains numerous processes, or projections.

The skull is divided into the neurocranium, housing the brain, and the viscerocranium, making up the face and jaw. The upper portion of the neurocranium makes up the calvaria, also known as the skullcap. The floor, or base, of the neurocranium consists of three fossae, or cavities, that house and protect soft tissue.

Bones of the skull are fused to neighboring bones at sutures; fusion is complete at approximately 20 years of age. Sutures are fibrous joints that are unique to the skull. The main sutures include the coronal suture, which fuses the frontal bone to two parietal bones, the sagittal suture, which fuses the parietal bones to each other, and the lambdoid suture, which fuses the occipital bone to the two parietal bones. The sutures represent weak spots of the skull that are susceptible to injury in trauma.

Figure Description

Anterior view of the skull

Key

A frontal bone
B nasal bone
C lesser wing of sphenoid bone
D greater wing of sphenoid bone
E parietal bone
F temporal bone
G ethmoid bone
H zygomatic bone
I vomer bone
J inferior nasal concha bone
K maxilla bone
L mandible
M supraorbital foramen
N optic canal
O superior orbital fissure
P inferior orbital fissure
Q zygomatic foramen
R infraorbital foramen
S nasal spine
T alveolar process
U mental foramen

The Skull

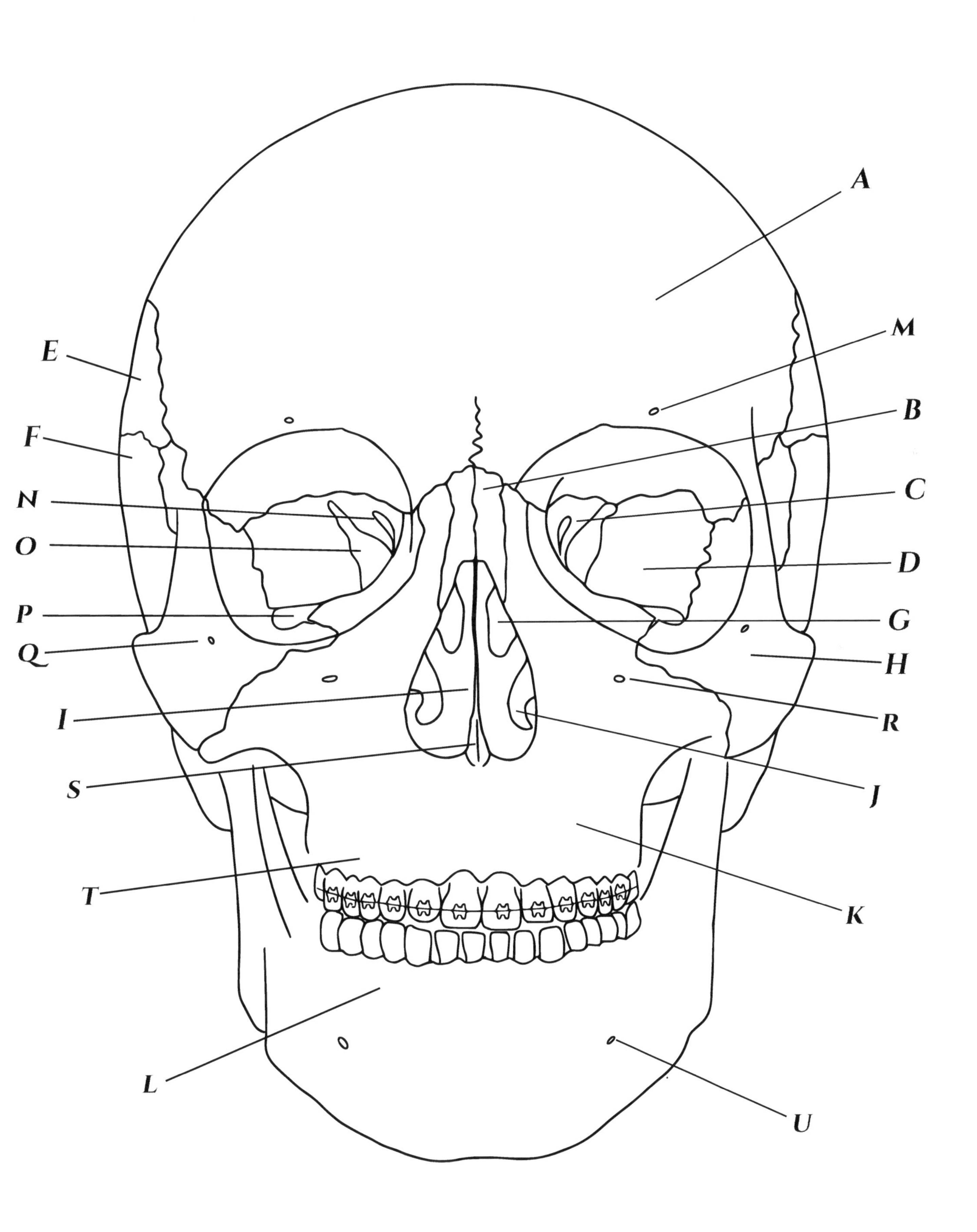

The Cranial Fossae

The floor of the cranial cavity consists of three major fossae, or depressions, designated as the anterior, middle, and posterior cranial fossae.

The anterior cranial fossa houses the frontal lobes of the cerebrum, and is the shallowest of the three fossae. The anterior cranial fossa also houses the olfactory tracts and bulbs. The crista galli and cribriform plate are found medially. The orbital plates of the frontal bone make up the floor of the anterior cranial fossa and the roof of the orbits.

The middle cranial fossa houses the temporal lobes of the cerebrum and the pituitary gland. The tuberculum sellae and sella turcica are found medially. The middle cranial fossa is bounded anteriorly by the posterior borders of the lesser wing of the sphenoid bone and posteriorly by the dorsum sellae of the sphenoid bone and petrous portion of the temporal bones. The optic canal and foramen rotundum, ovale, spinosum, and lacerum are found in the middle cranial fossa.

The posterior cranial fossa is the largest and deepest of the three fossae. It houses the cerebellum and occipital portion of the cerebrum, including the primary visual cortex. The foramen magnum is found medially, which is the large passageway for the spinal cord. Other foramina found in the posterior cranial fossa include the internal acoustic meatus and jugular foramen.

Figure Description

Superior view of the skull in a transverse plane with calvaria removed

Key

A anterior cranial fossa
B middle cranial fossa
C posterior cranial fossa
D frontal bone
E crista galli
F cribriform plate
G lesser wing of sphenoid bone
H hypophyseal fossa of sella turcica
I greater wing of sphenoid bone
J optic canal
K foramen rotundum
L foramen ovale
M foramen spinosum
N foramen lacerum
O internal acoustic meatus
P temporal bone
Q jugular foramen
R petrous ridge
S foramen magnum
T parietal bone
U occipital bone
V internal occipital ridge
W cerebellar fossa

THE CRANIAL FOSSAE

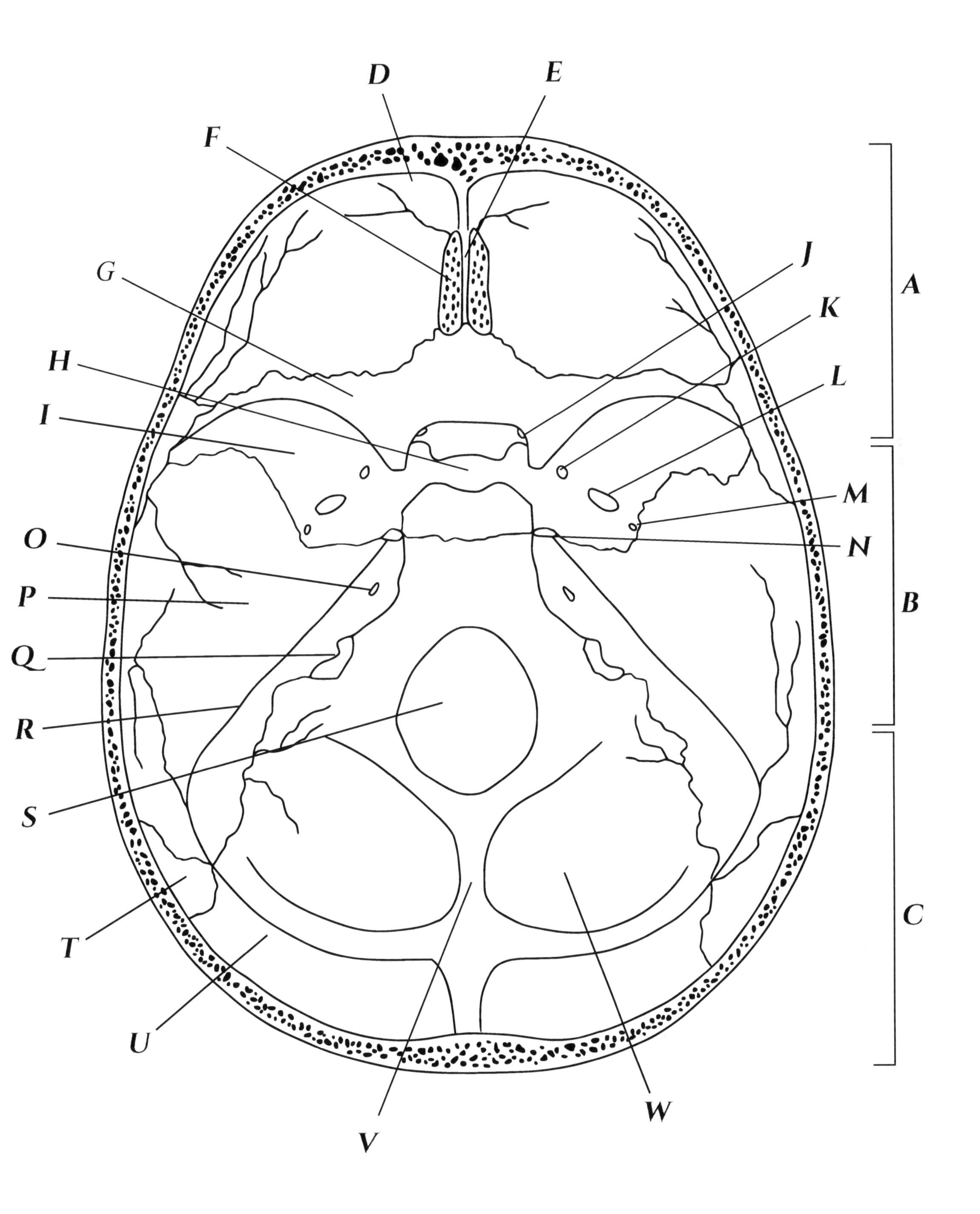

The Bony Orbit

The bony orbits of the skull form cavities in which the eyes are protected on all sides, except anteriorly. Seven bones contribute to each orbit, including the frontal, sphenoid, ethmoid, parietal, temporal, zygomatic, and lacrimal bones.

Each orbit contains an eye and optic nerve, as well as extraocular muscles, blood vessels, nerves, lymphatics, the lacrimal gland, the lacrimal sac, fascia, and adipose tissue. There are no empty spaces in the orbital cavities. The region of the orbit that is outside of the extraocular rectus muscles is extraconal, and the region within the extraocular rectus muscles surrounding the optic nerve is intraconal.

The orbit is pyramidal in shape, and is described as having a roof, lateral wall, floor, and medial wall. The medial walls of the two orbits are parallel, and the lateral walls form a 90 degree angle with each other. The medial wall consists of the lacrimal, ethmoid, maxilla, and sphenoid bones. The floor of the orbit consists of the maxilla, palatine, and zygomatic bones. The floor of the orbit is the roof of the maxillary sinus. The lateral wall consists of the zygomatic and sphenoid bones. The roof consists of the frontal and sphenoid bones. The anterior most rim of the bony orbit forms the orbital rim, which helps to protect the eye as it sits within the cavity.

The optic canal and superior and inferior orbital fissures are located at the apex of the orbit. The anterior and posterior ethmoidal foramina and the lacrimal foramen are found on the medial wall, and the recurrent meningeal foramen is on the lateral wall. The infraorbital groove runs along the floor of the orbit towards the infraorbital canal. Other orbital foramina include the zygomaticofacial canal laterally and the supraorbital foramen (or notch) superiorly.

Figure Description

Right bony orbit, anterior view

Key

A frontal bone
B supraorbital foramen
C lesser wing of sphenoid bone
D posterior ethmoidal foramen
E anterior ethmoidal foramen
F nasal bone
G optic canal
H parietal bone
I recurrent meningeal foramen
J superior orbital fissure
K greater wing of sphenoid bone
L temporal bone
M zygomatic bone
N zygomatico facial foramen
O lacrimal bone
P ethmoid bone
Q inferior orbital fissure
R infraorbital groove
S maxilla bone
T infraorbital foramen
U palatine bone

THE BONY ORBIT

anterior view

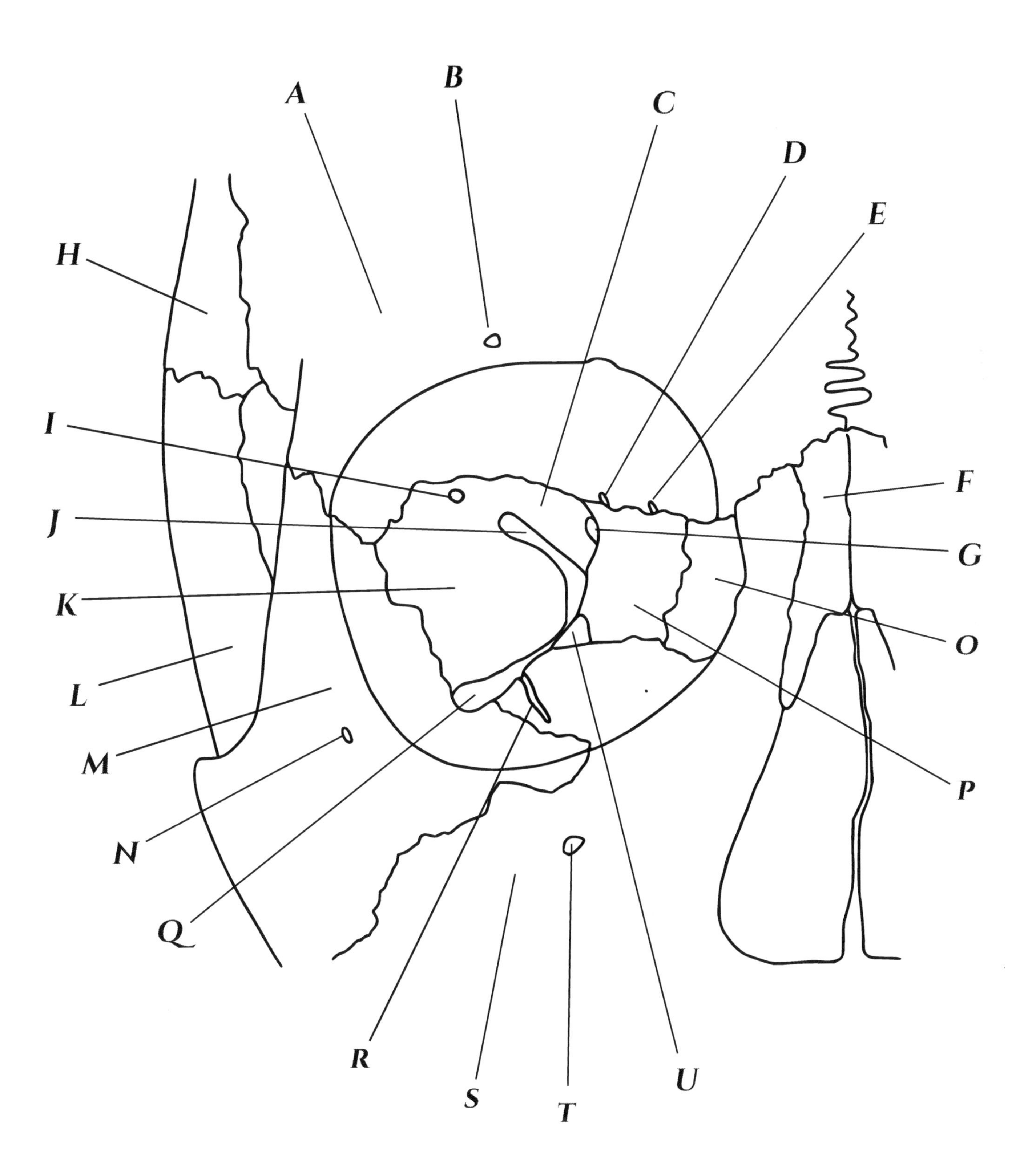

The Bony Orbit

The medial wall of the bony orbit consists of the lacrimal bone, orbital plate of the ethmoid bone, frontal process of the maxillary bone, and lesser wing of the sphenoid bone.

The lacrimal sac sits in a fossa in the anterior aspect of the medial wall. Tears pass from the lacrimal sac through the nasolacrimal duct, which exits the orbit through the lacrimal foramen and empties into the inferior meatus of the nasal cavity.

The anterior and posterior ethmoid foramina are also found on the medial wall of the orbit, which are passageways for the anterior and posterior ethmoid nerves and vasculature.

The medial wall of the orbit is the thinnest of the four walls. However, because it is stabilized by the horizontal plates of the ethmoidal air sinuses, the floor of the orbit is the weakest and most susceptible to orbital fracture in adults, followed by the medial wall. Children tend to have an increased risk of orbital roof fracture.

Figure Description

Medial wall of left bony orbit, lateral view with zygomatic bone removed

Key

A frontal sinus
B orbital plate of frontal bone
C anterior ethmoid foramina
D ethmoid bone
E posterior ethmoid foramina
F optic canal
G anterior clinoid process
H hypophyseal fossa
I posterior clinoid process
J nasal bone
K lacrimal sac fossa
L lacrimal bone
M lacrimal foramen
N palatine bone
O maxillary sinus
P maxilla bone

THE BONY ORBIT

medial wall

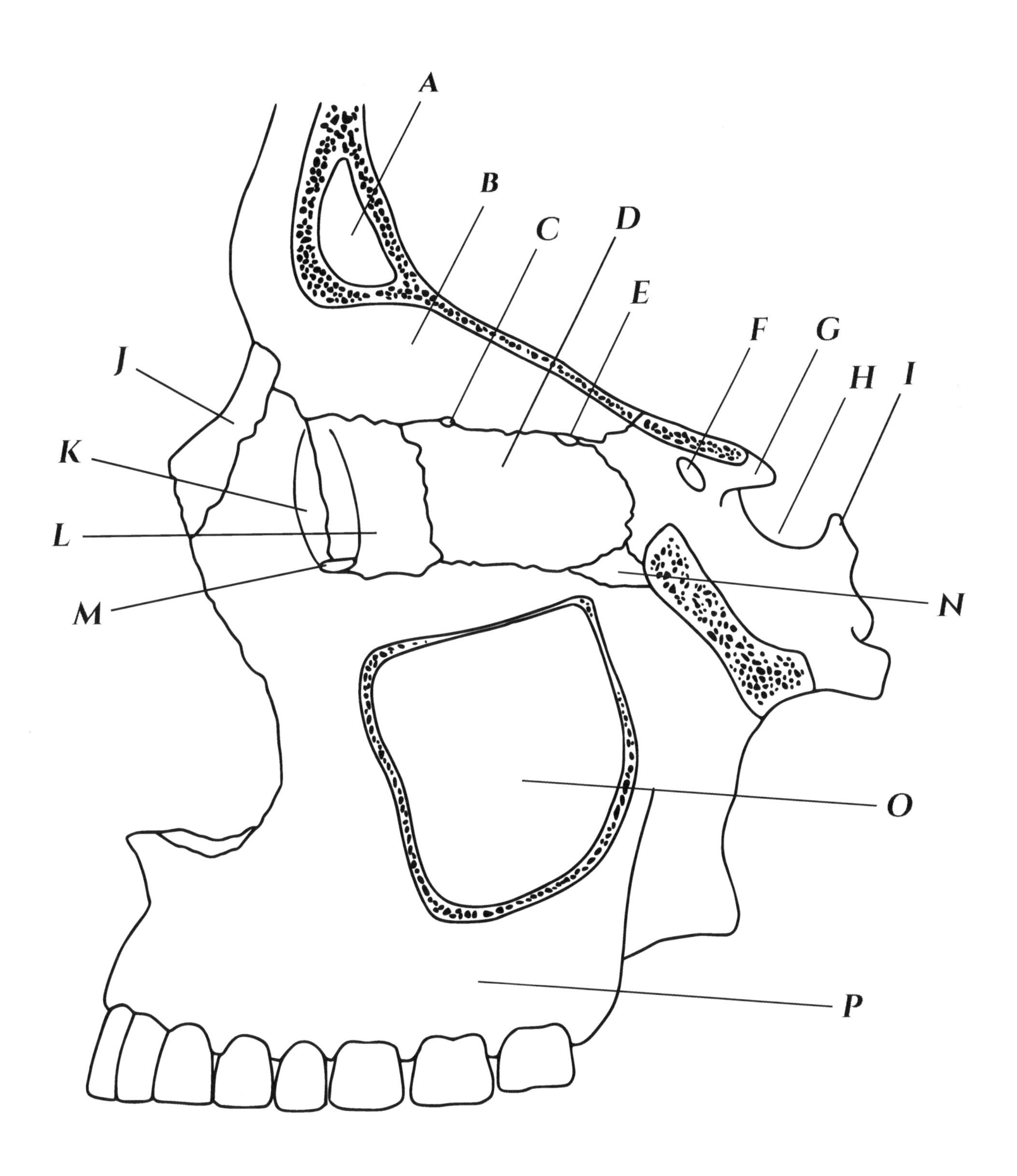

The Apex of the Orbit

The main passageways of nerves and vessels in and out of the orbit are the superior and inferior orbital fissures and optic canal, located at the apex of the bony orbit. Overlaying these passageways is the common tendinous ring, aka annulus of Zinn, which serves as the origin for the four rectus muscles. The levator palpebrae superioris and superior oblique muscles originate at the lesser wing of the sphenoid bone. The remaining extraocular muscle, the inferior oblique, originates on the medial wall of the orbit from the maxillary bone (not shown).

The optic nerve and ophthalmic artery pass through the optic canal within the common tendinous ring. Post-ganglionic sympathetic fibers also enter the orbit through the optic canal as a plexus surrounding the ophthalmic artery. The ophthalmic artery provides all of the blood supply to the eye.

Passing through the superior orbital fissure, outside of the common tendinous ring, are the recurrent meningeal artery, lacrimal and frontal nerves (of CN V1), the trochlear nerve (CN IV), and the superior ophthalmic vein. Within the common tendinous ring are the superior and inferior divisions of the oculomotor nerve (CN III), the nasociliary nerve, which is a branch of the ophthalmic nerve (CN V1), and the abducens nerve (CN VI).

Passing through the inferior orbital fissure are the zygomatic and infraorbital nerves, which are branches of the maxillary nerve (CN V2), the infraorbital artery and vein, and the inferior ophthalmic vein.

Figure Description

Right posterior orbit, highlighting the superior and inferior orbital fissures and annulus of Zinn

Key

A levator palpebrae superioris muscle
B superior rectus muscle
C superior oblique muscle
D medial rectus muscle
E inferior rectus muscle
F lateral rectus muscle
G optic nerve
H lacrimal nerve
I frontal nerve
J cranial nerve IV
K cranial nerve III, superior division
L nasociliary nerve
M cranial nerve VI
N cranial nerve III, inferior division
O zygomatic nerve
P infraorbital nerve
Q ophthalmic artery
R recurrent meningeal artery
S infraorbital artery
T infraorbital vein
U superior ophthalmic vein
V inferior ophthalmic vein
W optic canal
X superior orbital fissure
Y inferior orbital fissure
Z common tendinous ring

Apex of the Orbit
and annulus of Zinn

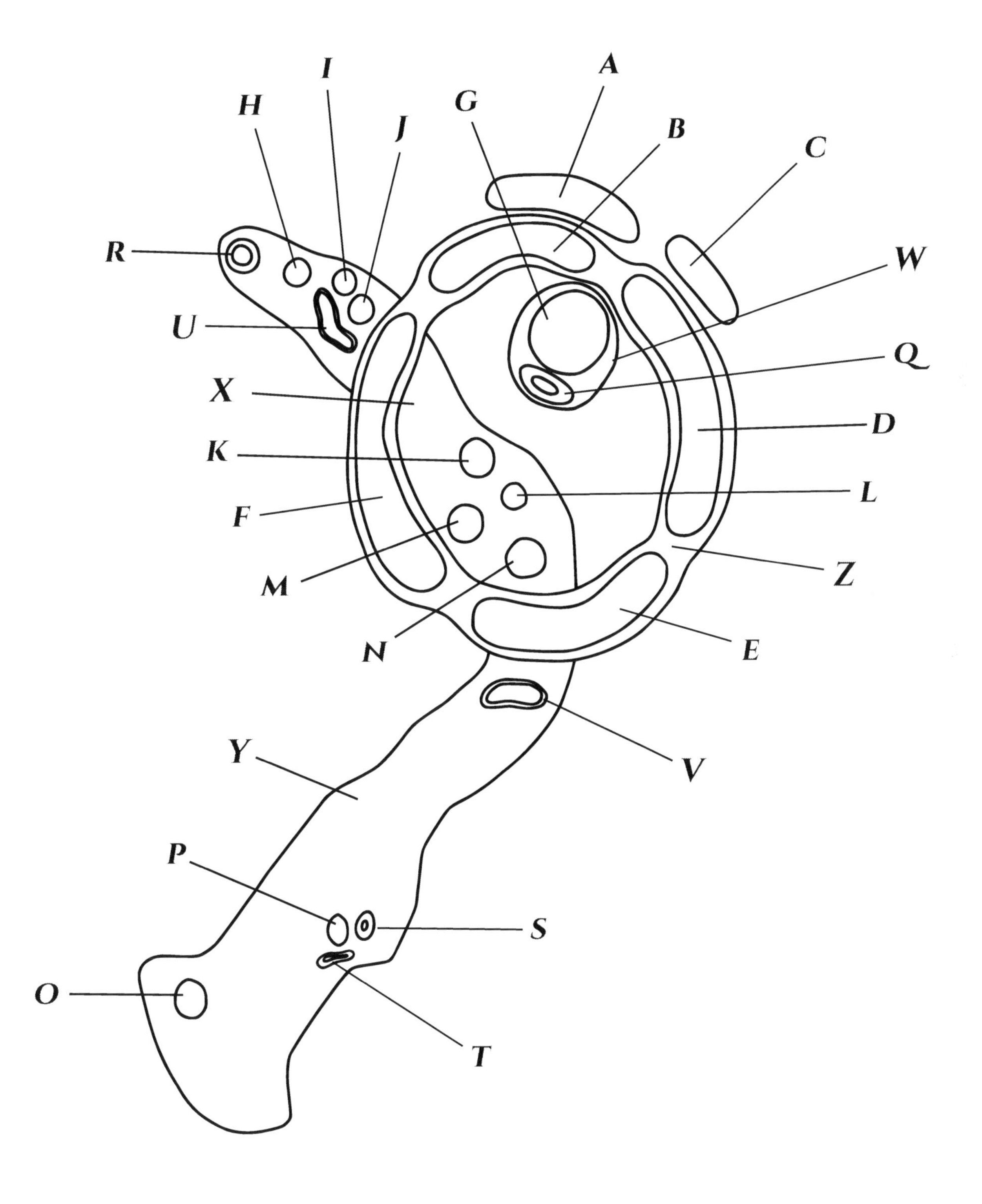

Paranasal Sinuses

The paranasal sinuses are air-filled spaces within the bones of the skull. There are four paired sinuses surrounding the nasal cavity, named after the bones in which they are located, including the frontal, ethmoid, maxillary, and sphenoid (not shown) sinuses.

The paranasal sinuses function to lighten the skull, protect soft tissue in injury, warm and humidify inhaled air, and resonate speech. The sinuses are lined with respiratory epithelium, a ciliated pseudostratified columnar epithelium, and connect to the nasal cavity via ostia, or small channels. The respiratory epithelium of the ostia is susceptible to inflammation, which can block the channels and cause a build up of mucous in the sinus cavities, leading to sinusitis.

Figure Description

Anterior view of face, overlaid with 3 of the 4 paranasal sinuses

Key

A frontal sinus
B ethmoid sinus
C maxillary sinus

Paranasal Sinuses

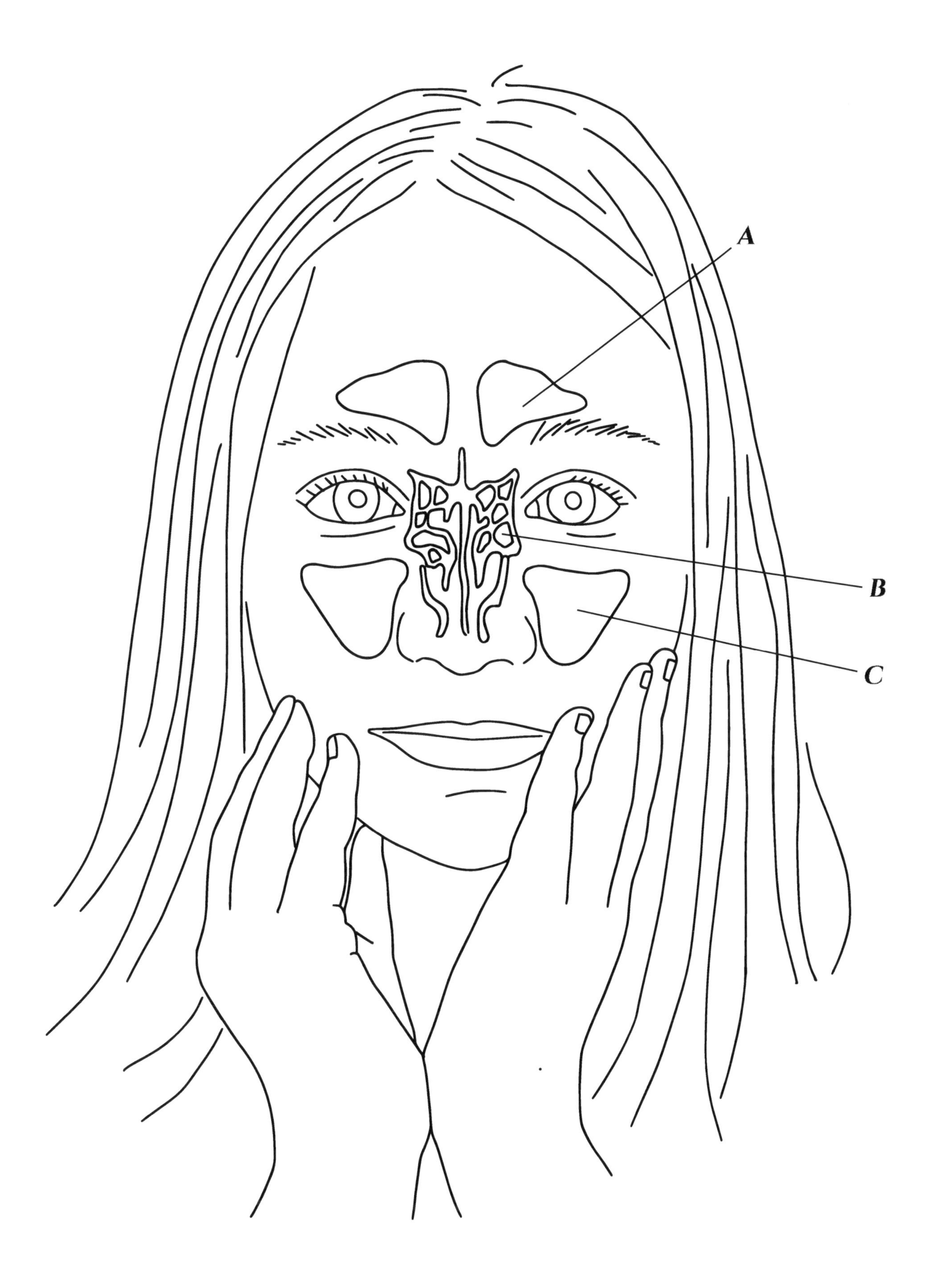

EXTRAOCULAR MUSCLES

There are seven striated muscles in each orbit. Six of them are termed extraocular muscles, which insert into the eye and function in eye movement. The seventh is the levator palpebrae superioris, which inserts into the eyelid and functions to elevate the superior eyelid. The extraocular muscles are striated voluntary muscles that are innervated by cranial nerves. They are derived from mesoderm in development. Their specific geometry, determined by the origin and insertion of each muscle, gives rise to each muscle's action, which may consist of up to three different directions of movement.

The four rectus muscles, superior, medial, inferior, and lateral, have their origin at the common tendinous ring, aka annulus of Zinn, at the orbital apex. The superior oblique muscle originates from the lesser wing of the sphenoid bone at the orbital apex, and the inferior oblique muscle originates from the maxillary bone on the medial wall of the orbit. The muscles have a tendinous insertion into the sclera of the eye. The four recti form the "muscle cone."

The superior and inferior rectus muscles are innervated by the oculomotor nerve (CN III), and insert into the globe at a 23 degree angle from the optical axis. The primary action of the superior rectus is elevation, the secondary action is intortion, and the tertiary action is adduction. The primary action of the inferior rectus muscle is depression, the secondary action is extortion, and the tertiary action is adduction.

The medial rectus muscle is innervated by the oculomotor nerve (CN III), and lateral rectus muscle is innervated by the abducens nerve (CN VI). The medial and lateral rectus muscles insert into the globe along the horizontal meridian and each has a single action. The medial rectus muscle adducts the eye (towards midline) and the lateral rectus muscle abducts the eye (away from midline).

The medial rectus muscle inserts closest to the limbus, at a distance of 5.5 mm. The inferior rectus inserts slightly farther at 6.5 mm, the lateral rectus at 6.9 mm, and the superior rectus at 7.7 mm. In an anterior view, drawing a line connecting these muscle insertion points forms a spiral, called the Spiral of Tillaux.

The superior oblique muscle is innervated by the trochlear nerve (CN IV), and the inferior oblique muscle is innervated by the oculomotor nerve (CN III). The superior oblique muscle passes from the apex of the orbit through the trochlea, or pulley, on the superior nasal aspect of the frontal bone. The superior oblique muscle becomes tendinous just before passing through the trochlea. The oblique muscles insert into the globe at a 54 degree angle. Therefore, the primary action of the superior oblique muscle is intortion, the secondary action is depression, and the tertiary action is abduction. The primary action of the inferior oblique muscle is extortion, the secondary action is elevation, and the tertiary action is abduction.

FIGURE DESCRIPTION

Right eye with six rotary extraocular muscles

KEY

A lateral view
B posterior view
C superior view
D superior rectus muscle
E superior oblique muscle (tendinous portion)
F trochlea of superior oblique muscle
G lateral rectus muscle
H inferior rectus muscle
I inferior oblique muscle
J medial rectus muscle
K optic nerve

EXTRAOCULAR MUSCLES

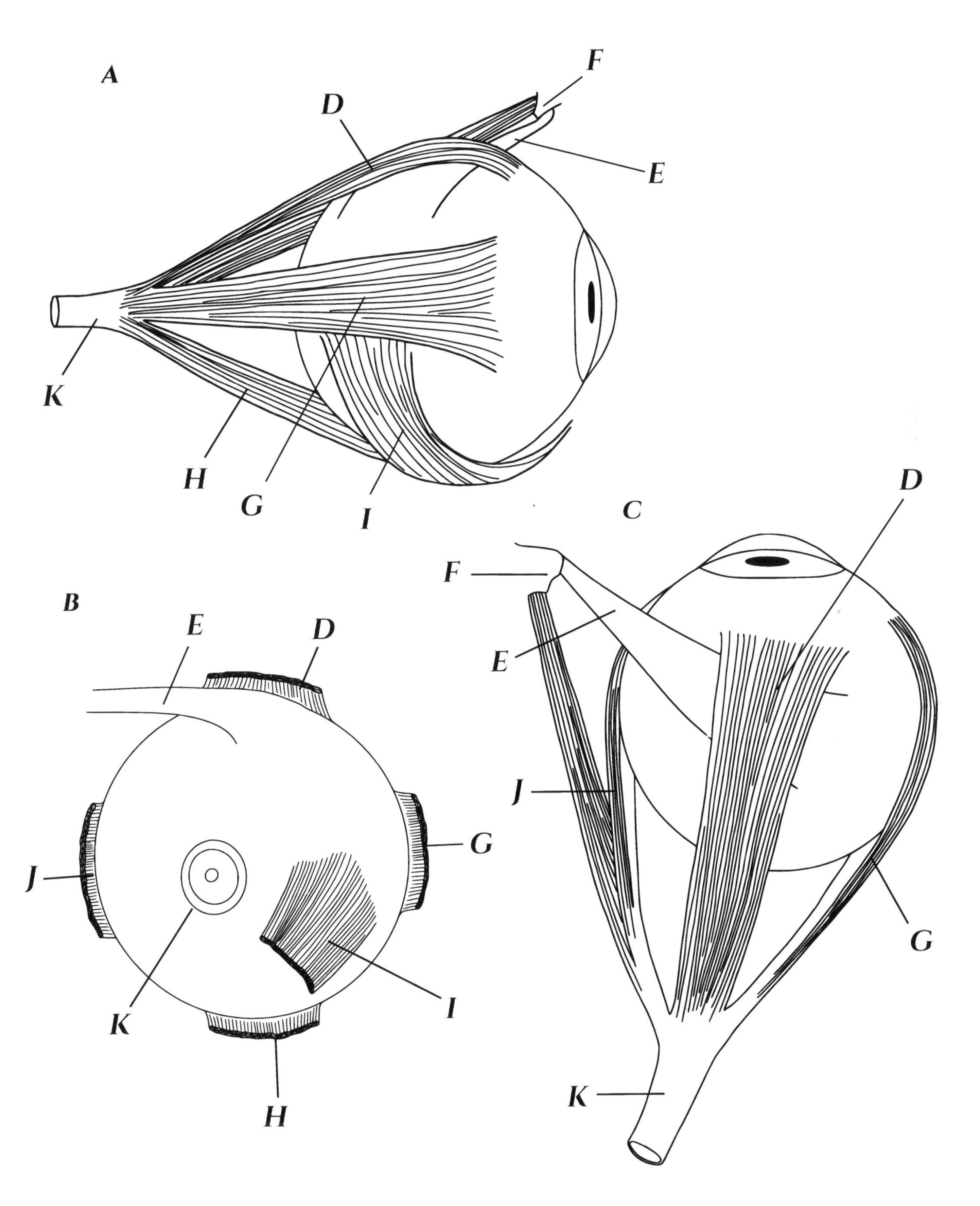

Extraocular Muscle Organization

The extraocular muscles are composed of striated skeletal muscle. Each muscle is surrounded by epimysium, which is a sheath of fibrous tissue. Within each muscle, the fascicles are surrounded by perimysium, and individual muscle fibers are surrounded by endomysium. Blood vessels, nerves, and fibroblasts can be found within the connective tissue layers.

Each of the six extraocular muscles that insert into the eye are organized into two distinct layers, orbital and global. The orbital layer of the extraocular muscles faces the bony orbit, and the global layer faces the eye. Muscle fibers of the orbital layer extend from the muscle origin to Tenon's capsule, whereas muscle fibers of the global layer extend the entire length of the muscle, from the origin to its insertion into the sclera.

The extraocular muscles are fatigue resistant and contain both singly innervated (fast twitch, more fatigue resistant) and multiply innervated (slow twitch, less fatigue resistant) muscle fibers. Singly innervated fibers are larger, whereas multiply innervated muscle fibers are smaller. Specific types of muscle fibers with variable amount of mitochondrial content are localized to either the orbital and global layers. Increased mitochondrial content is associated with increased fatigue resistance. In a histological cross section, mitochondria can be quantified by the number of oil droplets present.

The orbital layer has two types of muscle fibers - 1) singly innervated muscle fibers and 2) multiply innervated muscle fibers.

The global layer has larger muscle fibers than the orbital layer. There are four different types of muscle fibers in the global layer: 1) red singly innervated fibers, 2) intermediate singly innervated fibers, 3) pale (white) singly innervated fibers, and 4) multiply innervated fibers. Considering these three types of singly innervated muscle fibers, “red” fibers have the highest mitochondrial content and highest fatigue resistance, while “pale” fibers have the lowest mitochondrial content and lowest fatigue resistance.

Figure Description

Cross section of an extraocular muscle

Key

A orbital layer
B global layer
C muscle fibers
D epimysium
E perimysium
F blood vessel

Extraocular Muscle Organization

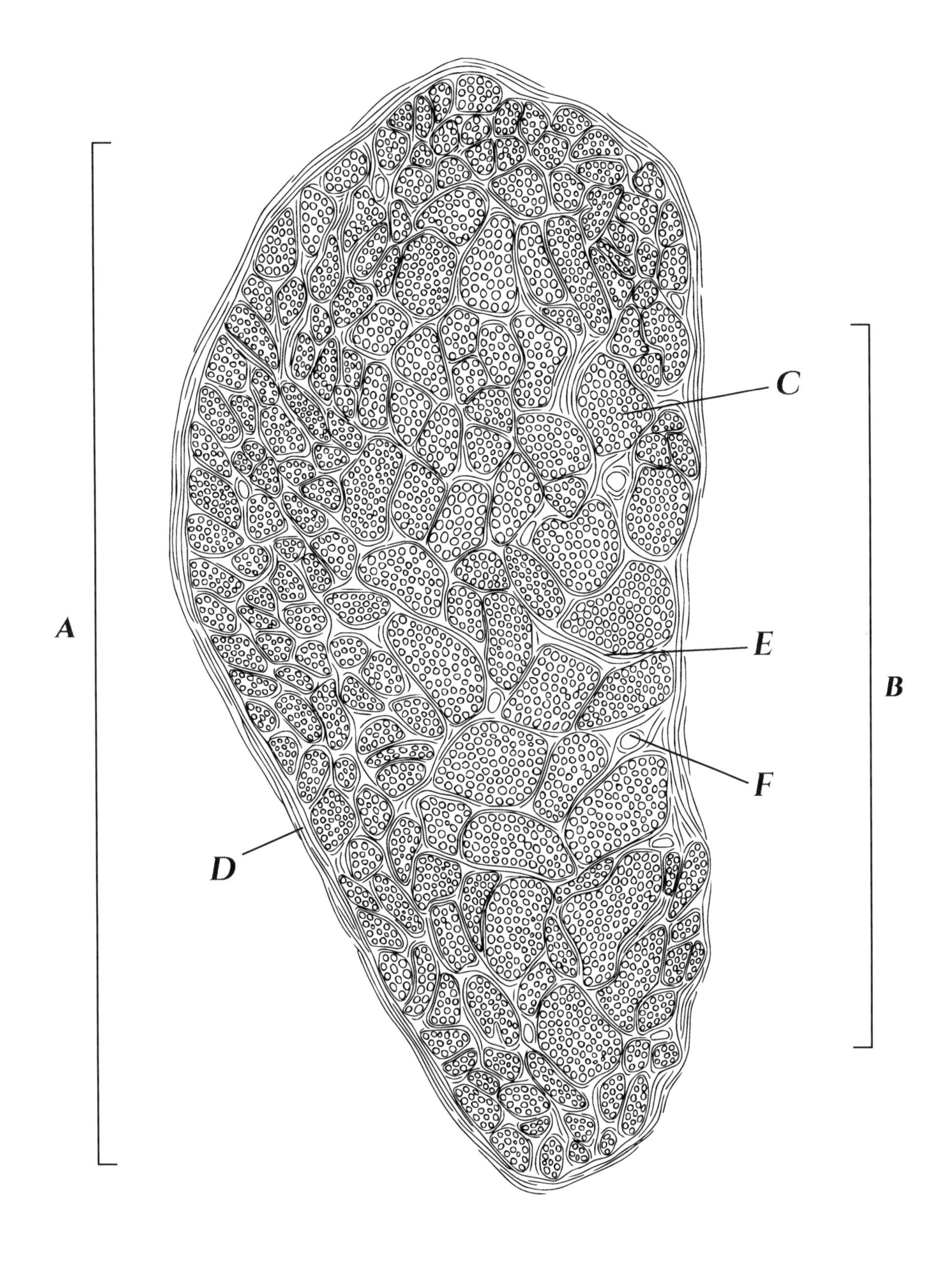

Extraocular Muscle Histology

Striated muscle consists of extrafusal fibers and intrafusal fibers. Extrafusal fibers are the main motor component of the muscle, and intrafusal fibers are the main sensory component, found within muscle spindles. Innervation to both types of fibers is via myelinated neurons that release acetylcholine at the neuromuscular junction.

Extrafusal muscle fibers are organized in sarcomeres, which are contractile units of actin and myosin filaments. Sarcomeres create the banded appearance visible in a histologic preparation. Extrafusal fibers are innervated by alpha motor neurons that contact the muscle at neuromuscular junctions. Nerve endings on singly innervated muscle fibers are en plaque, and on multiply innervated muscle fibers are en grappe.

Intrafusal fibers are located within the muscle spindle and consist of nuclear chain and nuclear bag fibers. Muscle spindles transmit information back to the brain about stretch, force, and proprioception. Intrafusal fibers are innervated by type 1a and 2 afferent neurons, as well as gamma efferent neurons. Nerve endings of type 1a neurons are annulospiral, and of type 2 neurons are flower spray. Nerve endings of gamma efferent neurons in the muscle spindle are plate and trail endings, found more distally on the intrafusal fibers.

Figure Description

Extraocular muscle histology

Key

A	extrafusal muscle fiber
B	sheath of muscle spindle
C	alpha efferent nerve to singly innervated fiber
D	en plaque nerve ending
E	alpha efferent nerve to multiply innervated fiber
F	en grappe nerve ending
G	group 1 afferent nerve
H	primary (annulospiral) nerve ending
I	group 2 afferent nerve
J	secondary (flower spray) nerve ending
K	gamma motor nerve to intrafusal muscle fibers
L	motor end plate nerve ending
M	intrafusal nuclear bag fiber
N	intrafusal nuclear chain fiber

Extraocular Muscle Histology

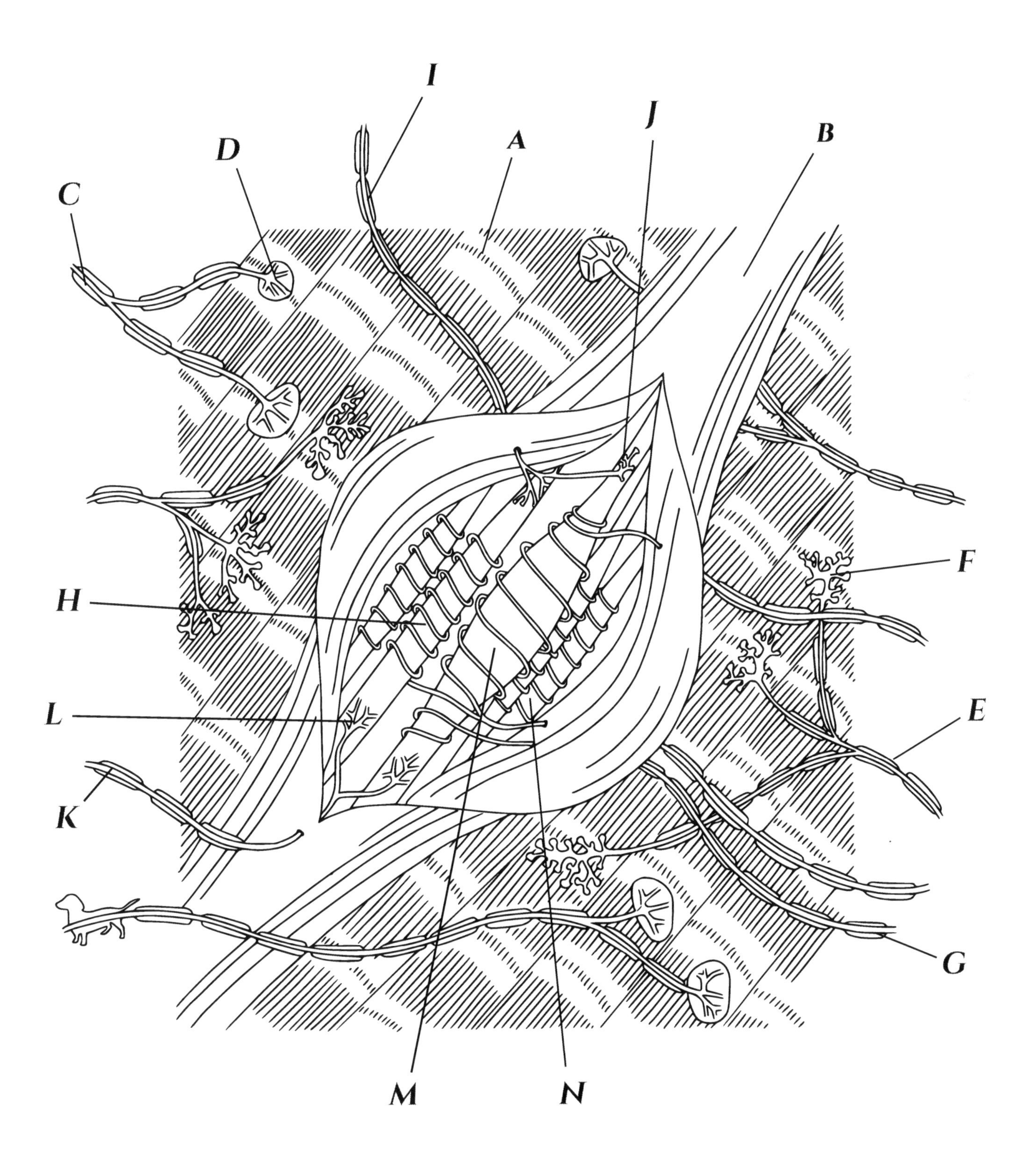

Ocular Surface Drainage

Tears are constantly produced, spread over the ocular surface, and drained. Numerous glands contribute to tears, including the tarsal (aka Meibomian) and Zeis glands, which contribute to the outermost sebaceous layer, the lacrimal and accessory lacrimal glands, which contribute to the middle aqueous layer, and the goblet cells, which contribute to the innermost mucin layer.

With each blink, tears are spread over the ocular surface. Tears collect along the inferior lid margin and toward the medial canthus, forming the lacus lacrimalis. The lacus lacrimalis accumulates due to gravity The tears will then leave the ocular surface through superior and inferior puncta, which are small holes surrounded by elastic connective tissue, known as the papilla lacrimalis. The papilla lacrimalis form an elevated region of the lid margin around the puncta, and serve to keep the puncta open. The puncta lead to canaliculi, consisting of a 2 mm vertical portion and 8 mm horizontal. The turn from the vertical to horizontal canaliculi is the ampulla. The superior and inferior horizontal canaliculi meet at the common canaliculus. The canaliculi empty to the lacrimal sac, which sits in the lacrimal sac fossa in the medial wall of the bony orbit. Several forces draw tears into the canaliculi, including gravity, blinking, and capillary action. With each blink, the canaliculi are pulled medially by Horner's muscle, which is the lacrimal portion of the orbicularis oculi muscle.

The lacrimal sac is approximately 10-12 mm in length, with a fundus superiorly and leading to the nasolacrimal duct inferiorly. The nasolacrimal duct passes through the lacrimal foramina of the bony orbit. The nasolacrimal duct is approximately 15-18 mm in length, and empties into the inferior meatus of the nasal cavity.

The drainage system contains a series of valves so that tears drain in one direction. The valves also prevent fluid and air from passing from the nasal cavity to the ocular surface. The valve at the junction of the common canaliculus and lacrimal sac is the valve of Rosenmuller. Valves along the nasolacrimal duct include the valve of Krause, spiral valve of Hyrtl, valve of Taillefer, and finally, the valve of Hasner, also known as the plica lacrimalis.

Figure Description

Right ocular drainage system

Key

A lacrimal gland, orbital portion
B lacrimal gland, palpebral portion
C superior vertical canaliculus
D superior ampulla
E superior horizontal canaliculus
F inferior vertical canaliculus
G inferior ampulla
H inferior horizontal canaliculus
I common canaliculus
J lacrimal sac
K valve of Rosenmuller
L nasolacrimal duct
M valve of Krause
N spiral valve of Hyrtl
O valve of Taillefer
P plica lacrimalis (valve of Hasner)
Q inferior meatus

Ocular Surface Drainage

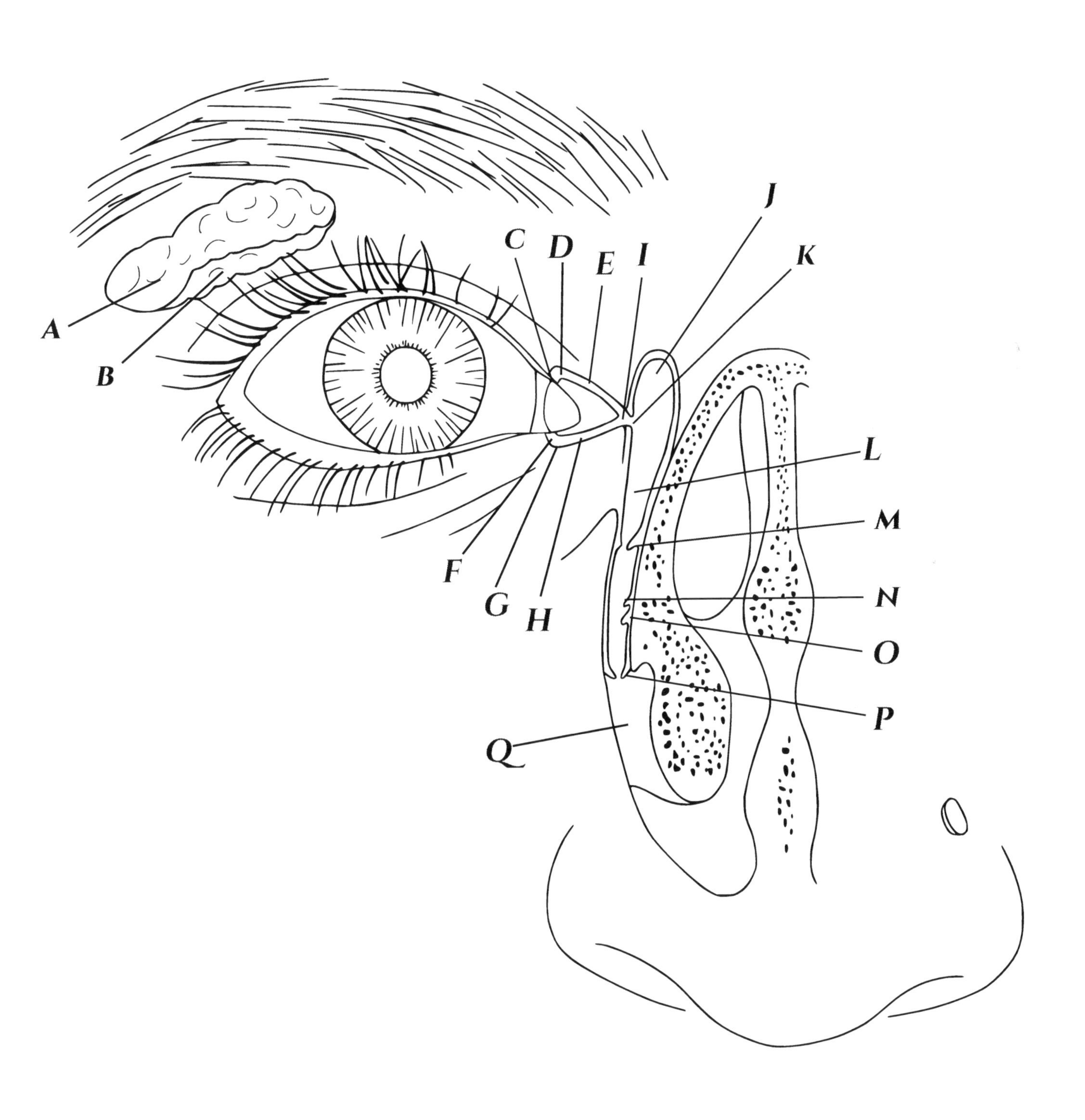

Lacrimal Gland Histology

The lacrimal gland is a serous exocrine gland, located in the lacrimal fossa of the frontal bone in the superior temporal bony orbit. The lacrimal gland is bilobed; it has two functionally similar regions, orbital and palpebral, that are separated by the aponeurosis of the levator palpebrae superioris muscle. The lacrimal gland is composed of acini with two types of secretory cells, one that secretes a serous product that contributes to the middle aqueous layer of the tear film, and a second that secretes mucin, which contributes to the inner layer of the tear film.

The lacrimal gland is a compound tubuloacinar gland comprised of lobules of acini separated by connective tissue, or interstitium. The acini are surrounded by myoepithelial cells, which are contractile cells that aid in secretion. Secretion is merocrine in nature. The cells of the acini release product into the lumen, which then follows as system of ducts to be secreted to the ocular surface at the superior fornix. The secretions of the lacrimal gland contain proteins and are bactericidal due to lysozyme and immunoglobulin content.

The interstitium surrounding the acini contains blood vessels and nerves, Blood supply is from the lacrimal artery, a branch of the ophthalmic artery. Blood drains to the lacrimal vein and to the superior ophthalmic vein. The lacrimal gland also contains lymphatic vessels that drain to the parotid lymph nodes.

Parasympathetic innervation stimulates secretion of the lacrimal gland. Parasympathetic innervation is provided by the facial nerve (CN VII). Pre-ganglionic fibers are carried by the greater petrosal nerve, a branch of the facial nerve. These fibers pass through the pterygoid canal and synapse in the pterygopalatine ganglion. Post-ganglionic fibers join the maxillary nerve (CN V2), enter the orbit as the rami lacrimalis, and travel to the lacrimal gland. Sensory innervation is from the lacrimal nerve of the ophthalmic division of the trigeminal nerve (CN V1).

Figure Description

Lacrimal gland histology

Key

A acinus
B secretory duct
C arteriole
D myoepithelial cell
E interstitium

Lacrimal Gland Histology

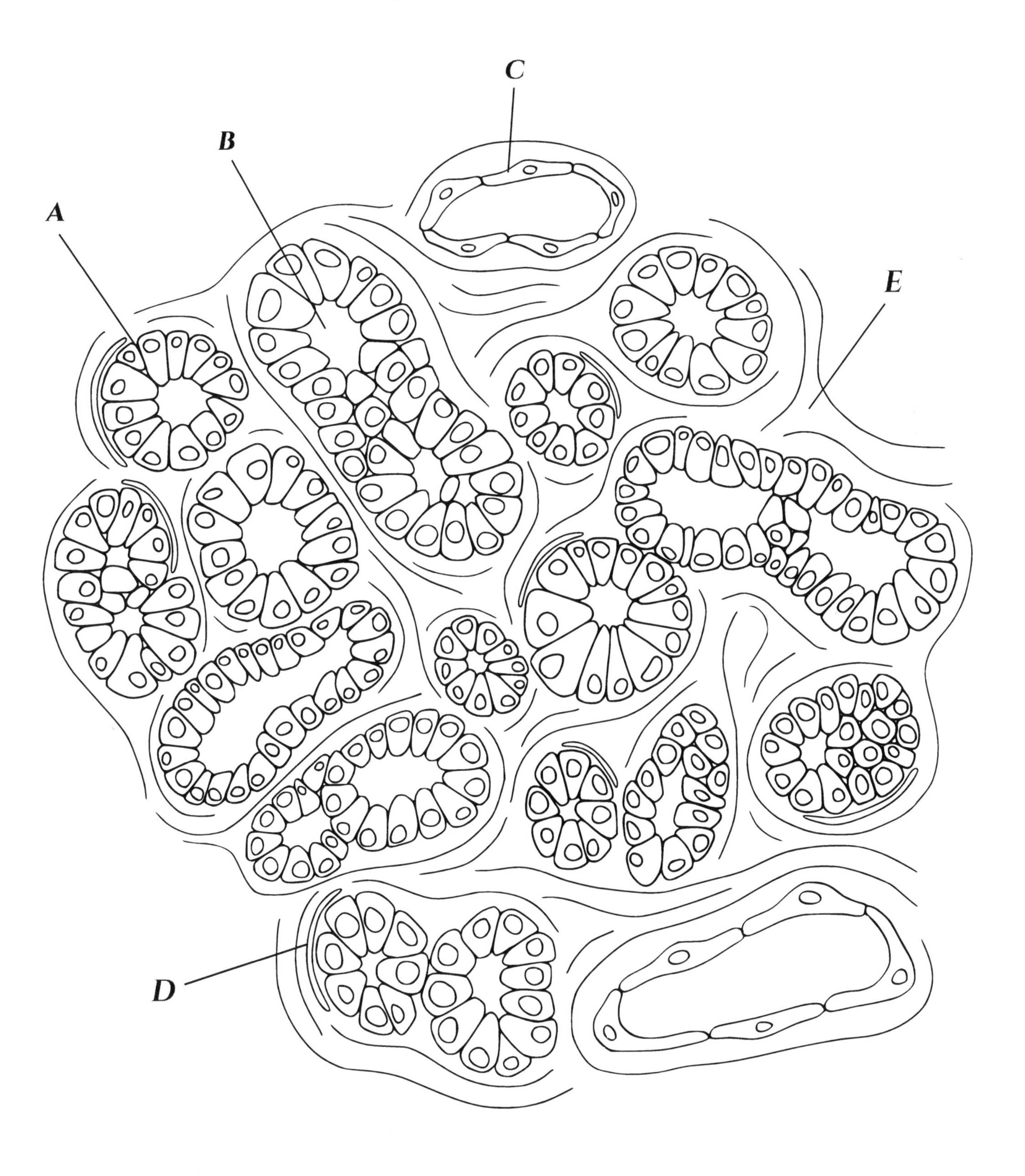

VASCULATURE

Blood Supply to the Eye and Orbit

The blood supply to the eye is provided by the ophthalmic artery, a branch of the internal carotid artery. The ophthalmic artery also provides some of the blood supply to the orbit, meninges, and structures of the nose and face.

After the internal carotid artery passes through the cavernous sinus, it gives off the ophthalmic artery, which enters the orbit through the optic canal. The ophthalmic artery travels within the muscle cone along the optic nerve and gives off numerous branches. The first branch of the ophthalmic artery is the central retinal artery, which, approximately 12 mm posterior to the globe, penetrates the optic nerve and travels anteriorly within the optic nerve to emerge from the optic nerve head and supply the inner retina.

The next branch is the lacrimal artery, which travels laterally and superiorly to supply and pass through the lacrimal gland. The lacrimal artery gives off the zygomatic artery posterior to the lacrimal gland, and the lateral palpebral artery anterior to the lacrimal gland.

The ophthalmic artery turns medially, passing over the optic nerve, giving off a lateral and medial posterior ciliary arteries, each of which gives rise to one long and several short posterior ciliary arteries. The long and short posterior ciliary arteries penetrate the sclera and supply the structures of the eye. As the ophthalmic artery passes over the optic nerve, it also gives off the muscular arteries, which follow the extraocular muscles anteriorly and give off the anterior ciliary arteries, which pierce the globe at the anterior segment.

The next branch is the supraorbital artery, which will exit the orbit through the superior orbital fissure (or notch) and supply the area around the eyebrow and forehead.

As the ophthalmic artery reaches the medial wall of the orbit, it gives off the ethmoidal arteries, which exit the orbit through the ethmoidal foramina.

Continuing anteriorly, the ophthalmic artery gives off the medial palpebral arteries, then terminates as the supratrochlear and dorsal nasal arteries.

Figure Description

Superior view of right orbit, roof of bony orbit removed, adapted from Gray, 1918

Key

A internal carotid artery
B ophthalmic artery
C central retinal artery
D lacrimal artery
E zygomatic artery
F lateral palpebral artery
G muscular artery (truncated)
H posterior ethmoidal artery
I anterior ethmoidal artery
J dorsal nasal artery
K supratrochlear artery
L medial palpebral artery
M supraorbital artery
N posterior ciliary arteries
O optic nerve

Blood Supply to the Eye and Orbit

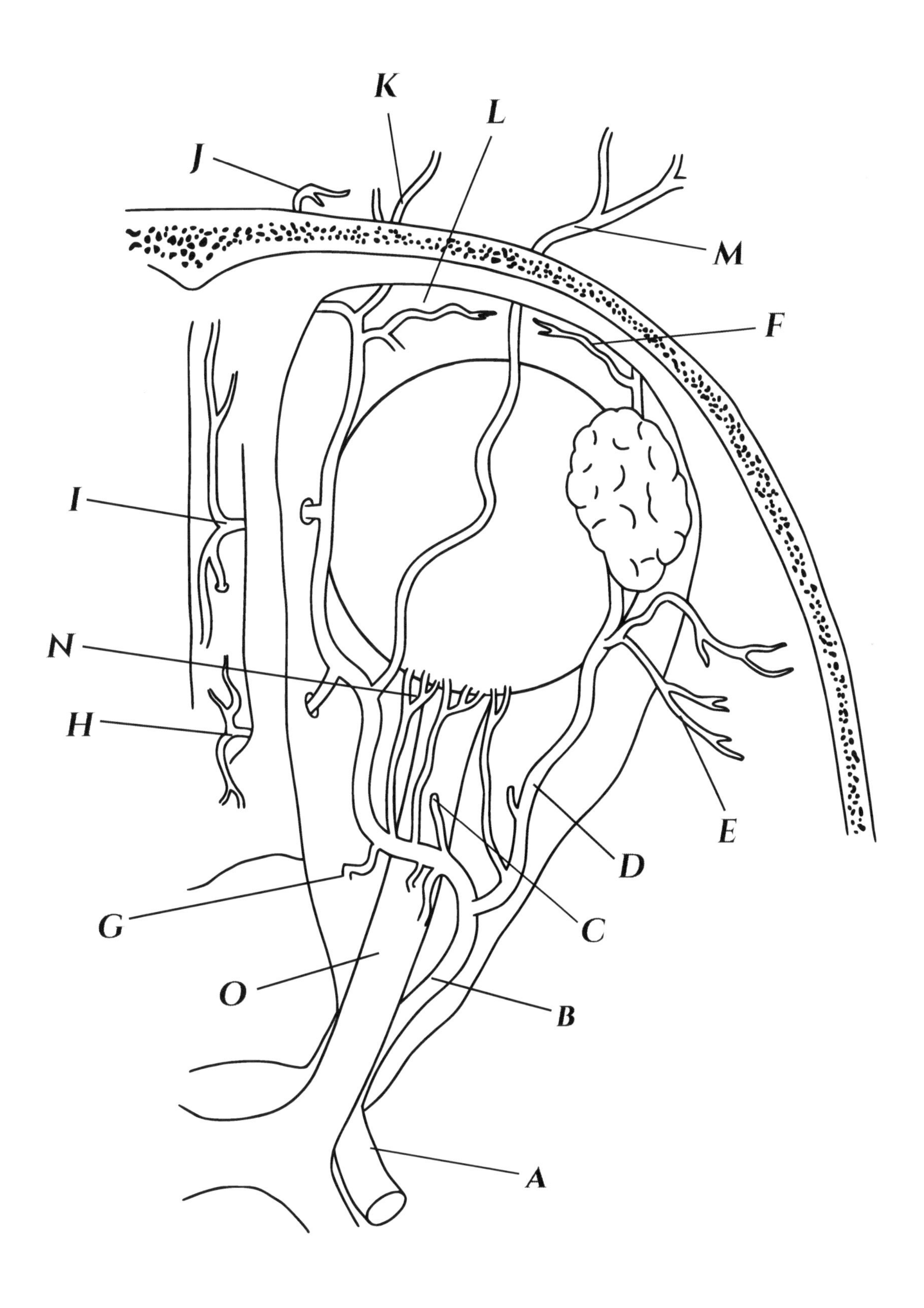

Anterior Ciliary Arteries

The anterior ciliary arteries provide the blood supply to the anterior portion of the eye. There are two anterior ciliary arteries associated with each rectus muscle, except for the lateral rectus, which has only one. The anterior ciliary arteries branch off of the muscular arteries, which are derived from the ophthalmic artery.

As the anterior ciliary arteries branch into episcleral arterioles, they travel within the episcleral towards the limbus. Some episcleral arterioles penetrate deeper into the eye, supplying the deep episclera, sclera, and ciliary body. The penetrating episcleral arterioles may contribute to the intramuscular circle within the ciliary muscle. It is also possible that these arterioles may contribute to the anterior choroid.

Other arterioles continue anteriorly in the episclera and anastomose with neighboring arterioles to form the episcleral circle. Some arterioles continue to the limbus to contribute to the marginal arcades that surround the cornea. Around the limbus, arterioles may travel towards the corneal margin, then turn 180 degrees, traveling within the conjunctiva; these vessels are recurrent arterioles.

Figure Description

Anterior view of right eye

Key

A lateral anterior ciliary artery
B superior anterior ciliary arteries
C medial anterior ciliary arteries
D inferior anterior ciliary arteries
E episcleral arteriole
F perforating episcleral arteriole
G episcleral circle
H marginal/limbal arterial arcades
I recurrent arteriole

Anterior Ciliary Arteries

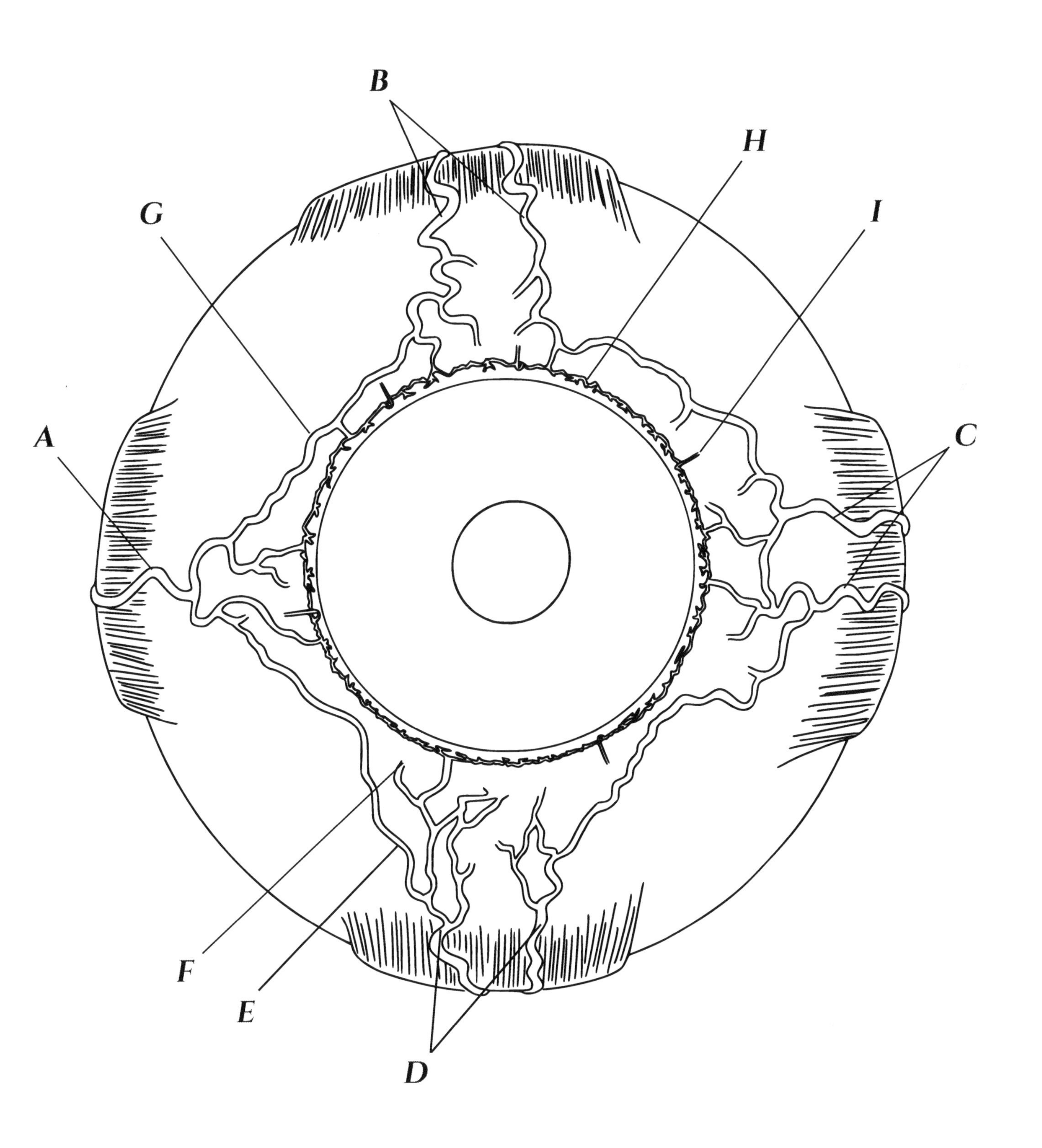

Long Posterior Ciliary Arteries

The long posterior ciliary arteries are branches of the posterior ciliary arteries, which branch from the ophthalmic artery. There are generally two long posterior ciliary arteries in each eye, which pierce the sclera on either side of the optic nerve and travel within the suprachoroidia to the anterior segment.

The long posterior ciliary arteries contribute blood supply to the iris, ciliary body, and anterior choroid. Within the ciliary body, branches of the long posterior ciliary arteries contribute to the major arterial circle of the iris, which is a discontinuous arterial circle. Branches from the major arterial circle of the iris travel radially towards the pupil and contribute to the minor arterial circle, which is found within the collarette of the iris.

Within the anterior segment of the eye, branches of the long posterior ciliary arteries form a vascular network with branches of the anterior ciliary arteries.

Figure Description

Superior view of right eye, outer tunic partially resected, highlighting the long posterior ciliary arteries

Key

A medial long posterior ciliary artery
B lateral long posterior ciliary artery
C major arterial circle
D minor arterial circle

Long Posterior Ciliary Arteries

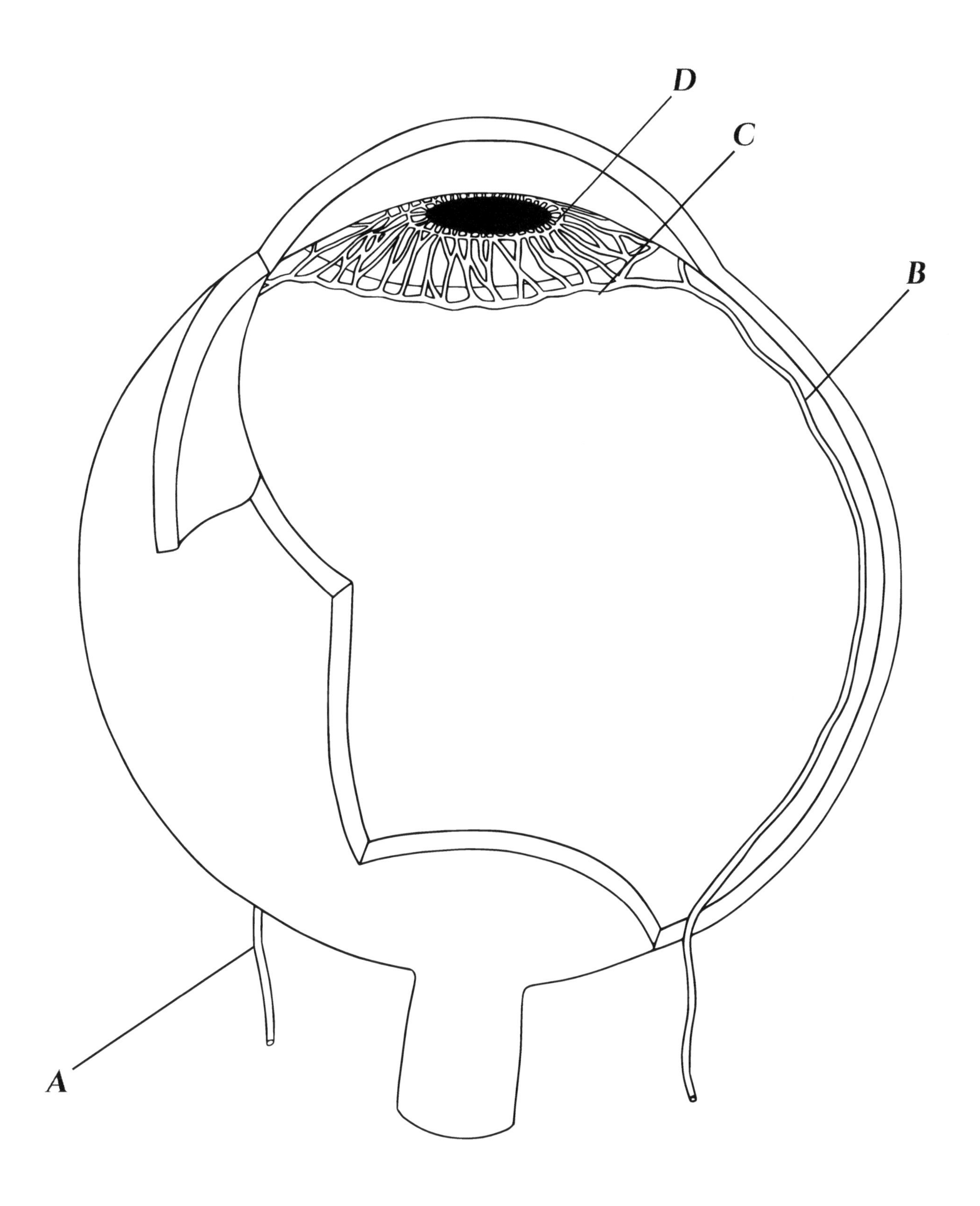

Short Posterior Ciliary Arteries

The short posterior ciliary arteries are branches of the posterior ciliary arteries, which branch from the ophthalmic artery. There are generally 10 to 20 short posterior ciliary arteries that pierce the sclera surrounding the optic nerve. The short posterior ciliary arteries contribute to the choroid vasculature, as well as the circle of Zinn-Haller, a vascular ring around the optic nerve within the sclera.

The short posterior ciliary arteries branch extensively within the choroid stroma and provide blood supply to the posterior segment. From the posterior segment, they travel anteriorly to just passed the equator of the eye.

Figure Description

Superior view of right eye, outer tunic partially resected, highlighting the short posterior ciliary arteries

Key

A short posterior ciliary arteries

Short Posterior Ciliary Arteries

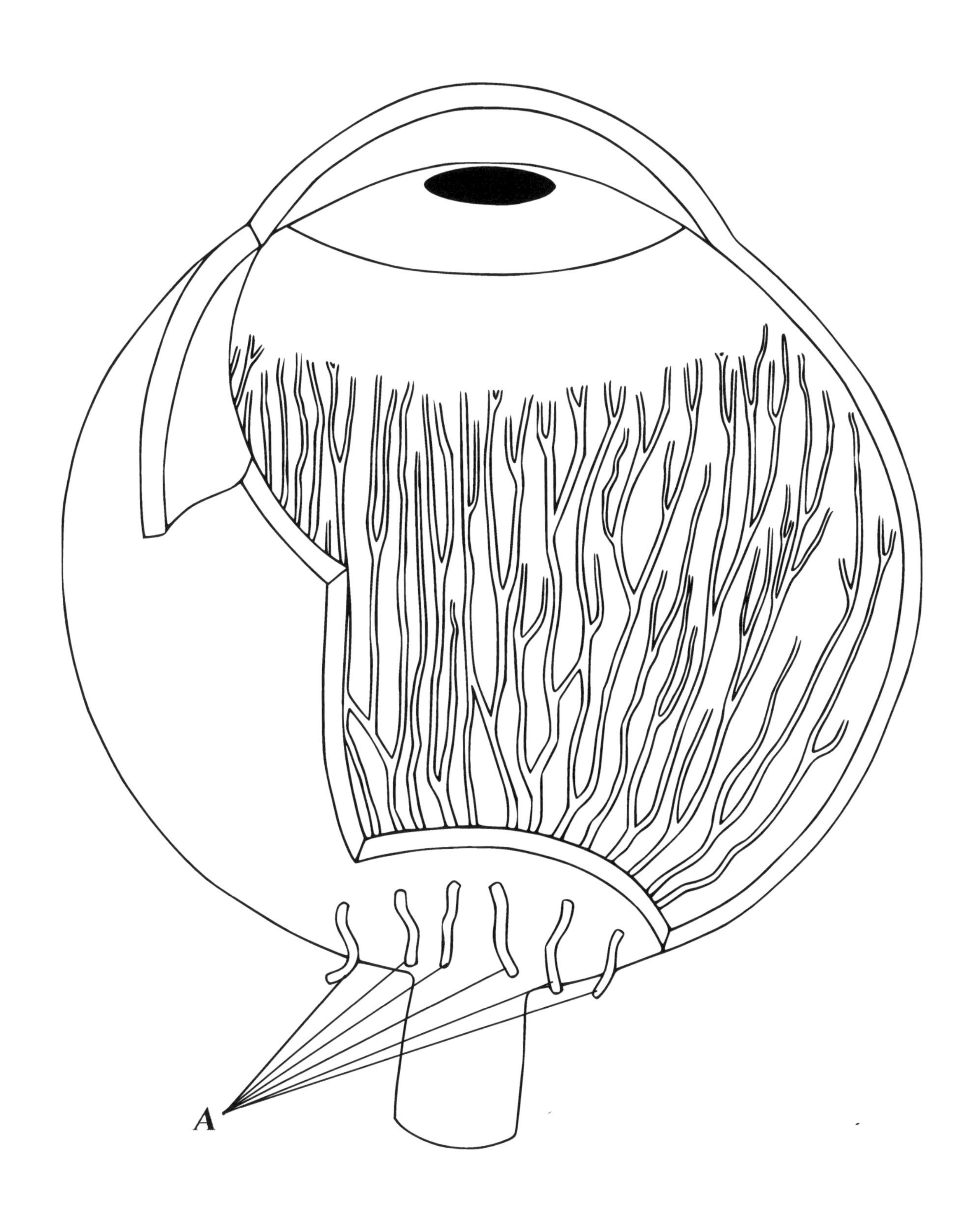

Vortex Veins

The vortex veins provide a route for venous drainage of the choroid, as well as most of the ciliary body and iris. There are generally four vortex veins per eye, one in each quadrant. The superior vortex veins drain into the superior ophthalmic vein to the cavernous sinus, and the inferior vortex veins drain into inferior ophthalmic vein and to the cavernous sinus and pterygoid plexus. There may be collateral circulation between the superior and inferior ophthalmic veins.

Figure Description

Superior view of right eye, outer tunic partially resected, highlighting the vortex veins, adapted from Cunningham, 1918

Key

A superior nasal vortex vein
B superior temporal vortex vein

VORTEX VEINS

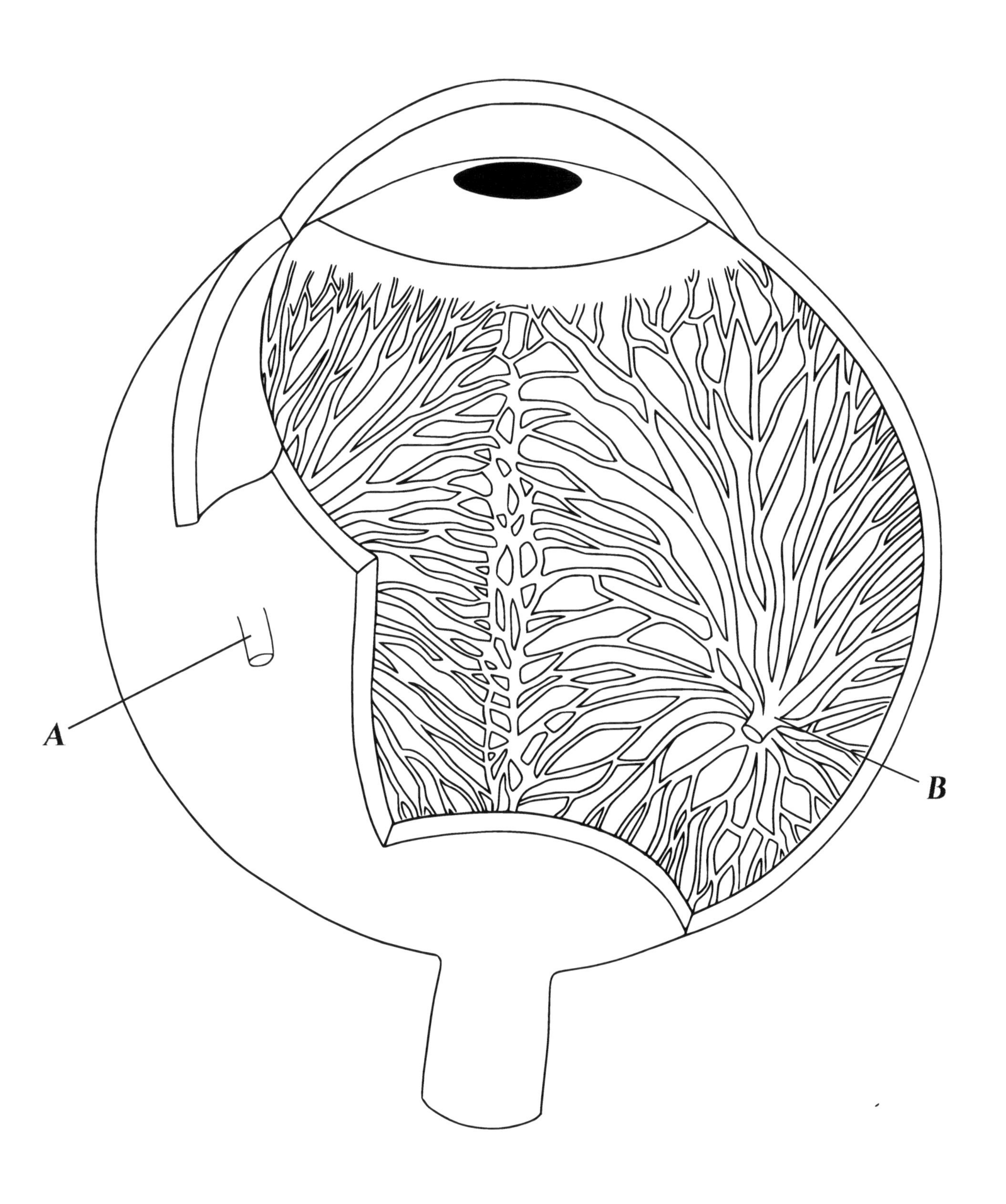

Venous Drainage from the Eye and Orbit

Blood drains from the eye via the central retinal vein and vortex veins, which drain posteriorly to the superior and inferior ophthalmic veins. The lacrimal, ethmoidal, and muscular veins of the orbit also drain into the ophthalmic veins.

The superior ophthalmic vein exits the orbit through the superior orbital fissure, and the inferior ophthalmic vein exits the orbit through the inferior orbital fissure, both outside of the annulus of Zinn. The superior ophthalmic vein drains into the cavernous sinus, which drains to the superior and inferior petrosal sinuses to the sigmoid sinus and to the internal jugular vein. The inferior ophthalmic vein drains to the cavernous sinus, as well as to the pterygoid plexus, which drains to the maxillary vein, to the retromandibular vein, and then to the external and internal jugular veins.

The veins of the skin around the orbits, including the eyelids, generally follow the arterial supply. These include the supraorbital, supratrochlear, and superior and inferior palpebral veins, which drain to the angular vein to the facial vein, then to the internal jugular vein.

Figure Description

Left bony orbit, lateral view with zygomatic bone removed

Key

A supraorbital vein
B supratrochlear vein
C superior palpebral vein
D inferior palpebral vein
E angular vein
F infraorbital vein
G facial vein
H lacrimal vein
I superior temporal vortex vein
J inferior temporal vortex vein
K superior ophthalmic vein
L central retinal vein
M inferior ophthalmic vein
N ophthalmic vein
O cavernous sinus
P emissary veins
Q pterygoid plexus
R maxillary vein
S retromandibular vein
T internal jugular vein

Venous Drainage from the Eye and Orbit

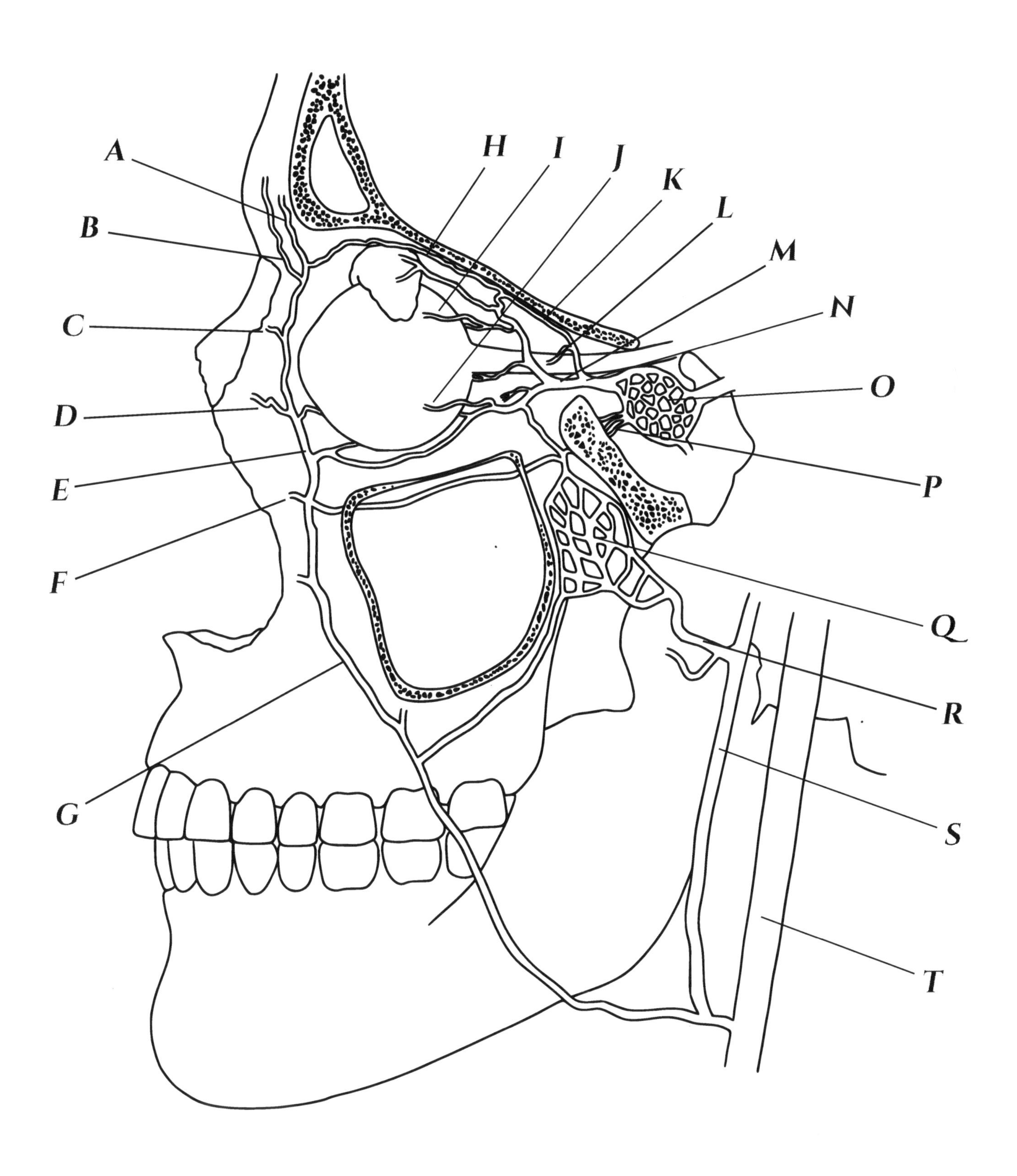

Cavernous Sinus

The cavernous sinus is a paired blood-filled dural sinus overlying the body of the sphenoid bone. It is situated anterior to the brainstem and posterior to the orbit; therefore, structures passing between the brain and orbit traverse through the cavernous sinus, including the oculomotor nerve (CN III), trochlear nerve (CN IV), ophthalmic nerve (CN V1), maxillary nerve (CN V2), and abducens nerve (CN VI). The internal carotid artery, specifically the carotid siphon, also passes through the cavernous sinus. The optic chiasm is superior to the cavernous sinus, and the pituitary gland lies within the sella turcica of the sphenoid bone, between the pair of cavernous sinuses.

The cavernous sinus receives blood from the superior and inferior ophthalmic veins, as well as from the superficial middle, deep, and inferior cerebral veins. The cavernous sinus drains posteriorly to the superior and inferior petrosal sinuses, and inferiorly to the pterygoid plexus. These systems ultimately converge on the internal jugular vein.

Figure Description

Coronal section of brain and cavernous sinus

Key

A optic chiasm
B pituitary gland
C extracavernous internal carotid artery
D CN III, oculomotor nerve
E CN IV, trochlear nerve
F intracavernous internal carotid artery
G CN V1, ophthalmic nerve
H CN VI, abducens nerve
I CN V2, maxillary nerve
J sphenoid sinus
K sphenoid bone
L nasopharynx

Cavernous Sinus

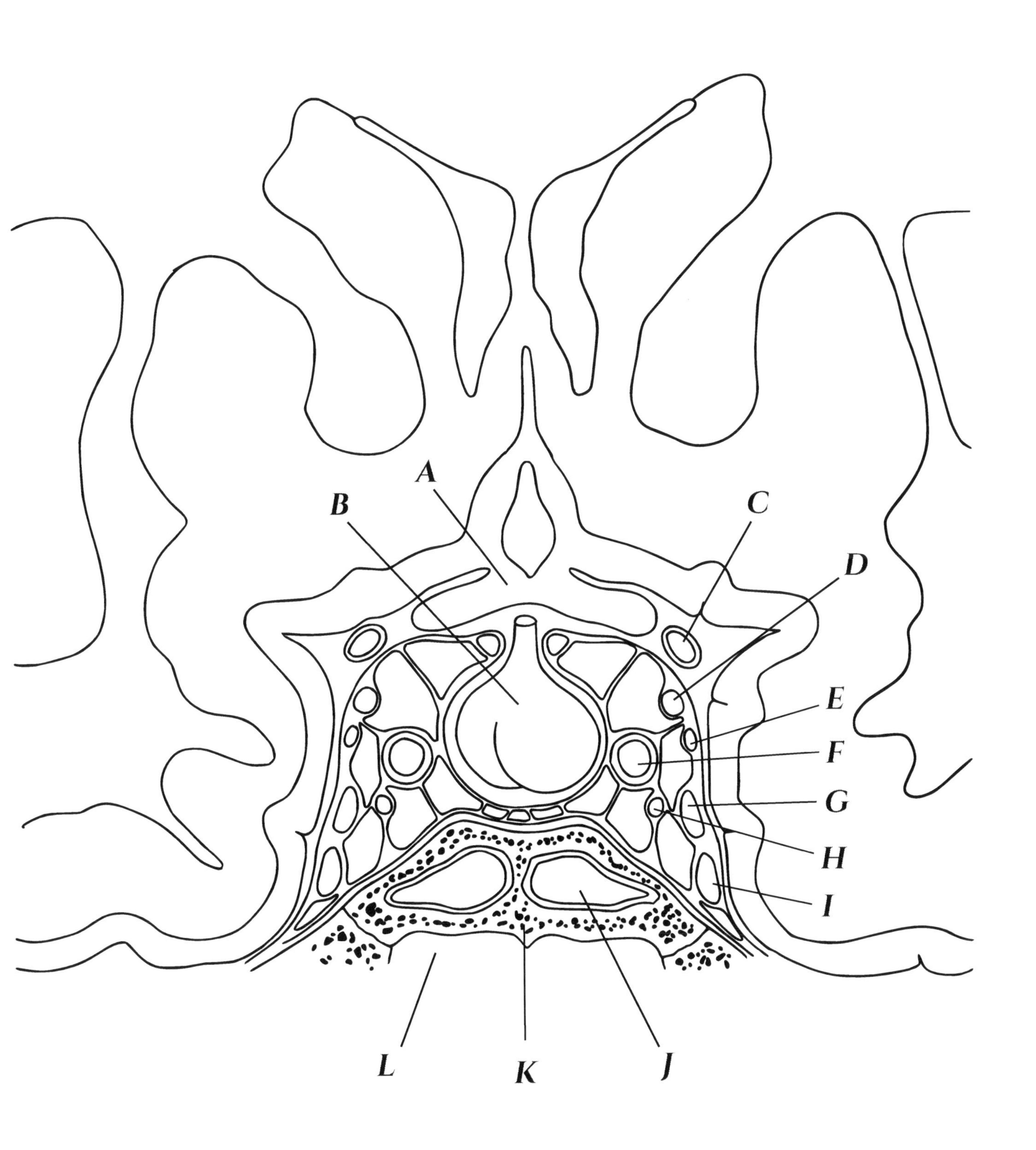

THE EYELID

Eyelid Arterial Supply

The eyelid has a rich arterial supply that is primarily provided by branches of the ophthalmic artery, which is a branch of the internal carotid artery. The temporal and facial arteries, which are branches of the external carotid artery, also contribute to the arterial supply of the eyelid.

The palpebral portion of the eyelid is supplied by branches of the medial and lateral palpebral arteries. The medial palpebral artery is a branch of the ophthalmic artery, and divides into superior and inferior medial palpebral arteries. The lateral palpebral artery is a branch of the lacrimal artery, which is a branch of the ophthalmic artery. The lateral palpebral artery also branches into superior and inferior lateral palpebral arteries. Branches of the medial and lateral palpebral arteries arch over the eyelids and anastomose, forming two arcades each on the upper and lower eyelids. On the upper eyelid, these are the superior peripheral and superior marginal arcades, and on the lower eyelid, the inferior marginal and inferior peripheral arcades. The arcades travel in the submuscular areolar layer of the eyelid.

The superior orbital region of the eyelid is supplied by the dorsal nasal, supratrochlear, and supraorbital arteries medially, and by the frontal branch of the temporal artery laterally. The inferior orbital region of the eyelid is supplied by the angular, facial, and infraorbital arteries medially, and the facial and zygomaticofacial arteries laterally.

Figure Description

Right orbit with overlying arteries

Key

A	frontal branch of temporal artery
B	lacrimal artery
C	superior lateral palpebral artery
D	inferior lateral palpebral artery
E	zygomaticofacial artery
F	superficial temporal artery
G	transverse facial artery
H	inferior marginal arterial arcade
I	inferior peripheral arterial arcade
J	supraorbital artery
K	supratrochlear artery
L	dorsal nasal artery
M	superior medial palpebral artery
N	angular artery
O	inferior medial palpebral artery
P	facial artery
Q	infraorbital artery
R	superficial temporal artery
S	superior peripheral arcade
T	superior marginal arcade

Eyelid Arterial Supply

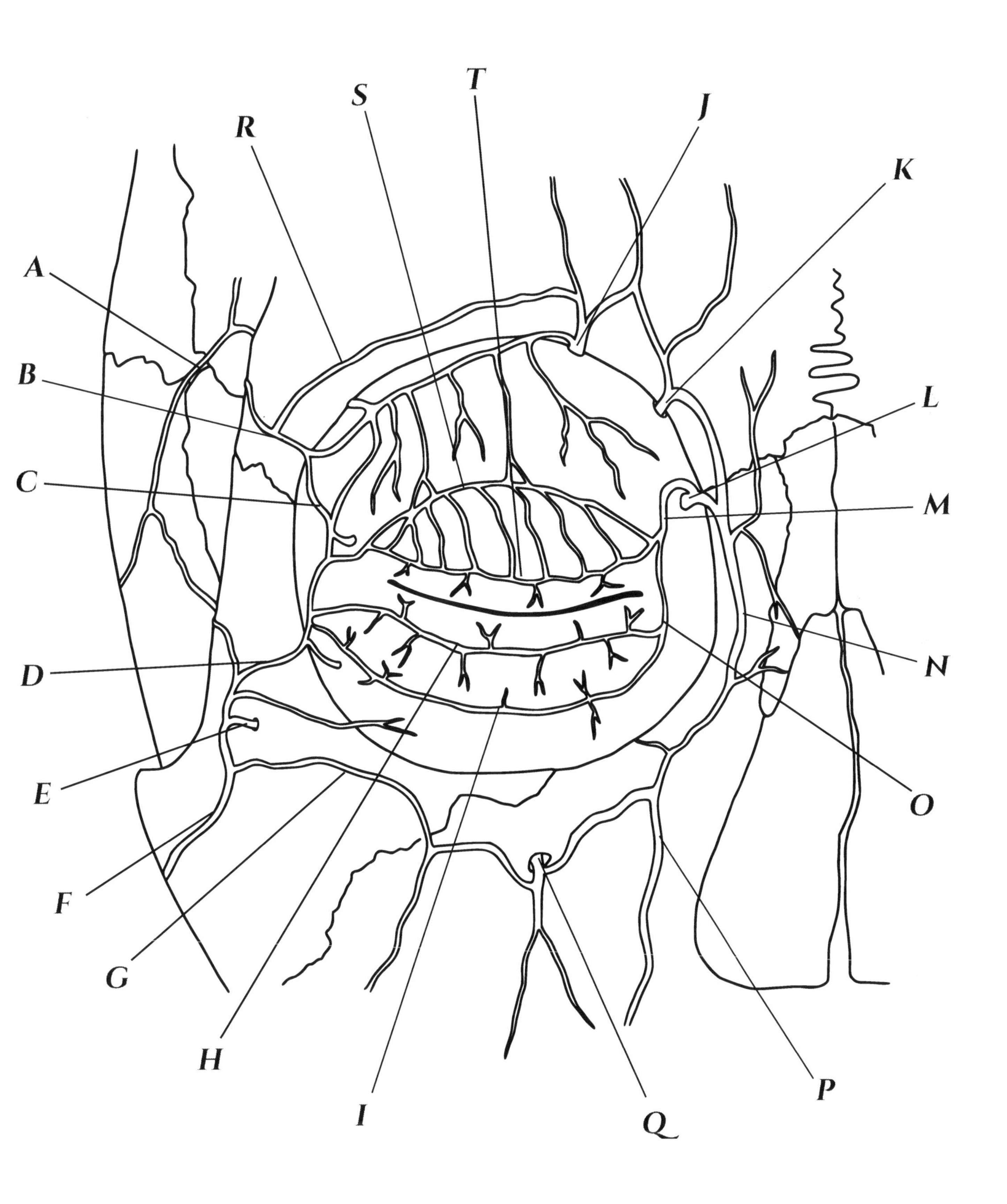

Eyelid Venous Drainage

The venous drainage from the eyelid roughly parallels the arterial supply, primarily draining to the anterior facial and superficial temporal veins. The palpebral region of the eyelid drains to the superior (upper eyelid) and inferior (lower eyelid) palpebral veins, which drain into the medial and lateral palpebral veins. The superior orbital region of the eyelid drains to the nasofrontal, supratrochlear, and supraorbital veins, and the inferior orbital region drains to the superficial, infraorbital, and angular veins.

The superficial temporal vein joins the maxillary vein, then continues as the retromandibular vein. The facial vein also joins the retromandibular vein. The retromandibular vein drains into the external and internal jugular veins.

Figure Description

Right orbit with overlying veins

Key

A	lateral palpebral vein
B	superficial temporal vein
C	inferior peripheral vein
D	superior palpebral vein
E	superior peripheral vein
F	supraorbital vein
G	supratrochlear vein
H	nasofrontal vein
I	dorsal nasal vein
J	medial palpebral vein
K	angular vein
L	infraorbital vein
M	facial vein

Eyelid Venous Drainage

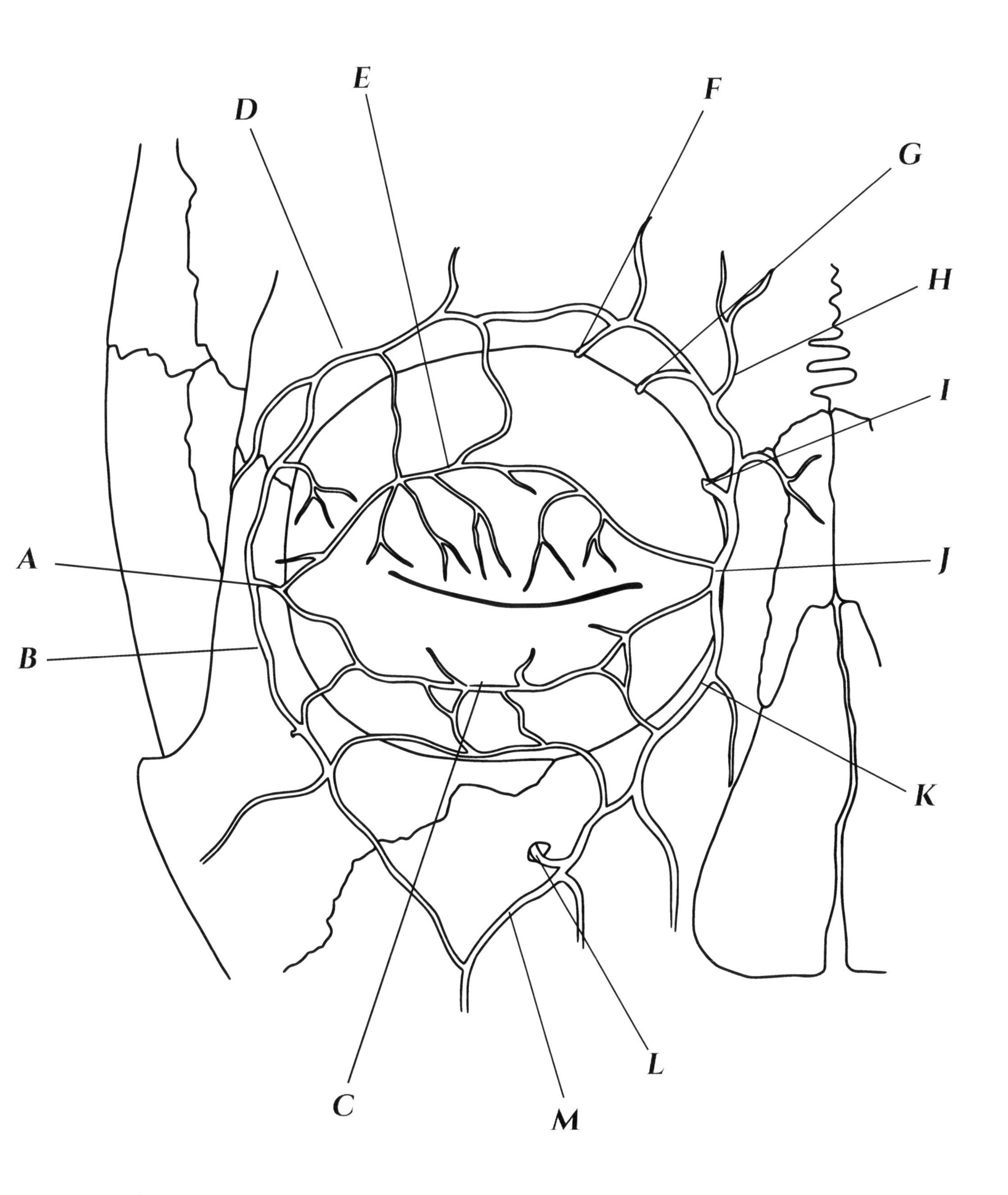

Eyelid Sensory Innervation

Sensory innervation of the eyelid is provided by the ophthalmic and maxillary divisions of the trigeminal nerve (CN V). As the ophthalmic nerve (CN V1) approaches the orbit, it divides into three branches, the lacrimal, frontal, and nasociliary nerves. The lacrimal and frontal nerves enter the orbit through the superior orbital fissure, outside of the annulus of Zinn. The nasociliary nerve enters the orbit through the superior orbital fissure, inside the annulus of Zinn.

The lacrimal nerve is the smallest branch of the ophthalmic nerve. It travels along the lateral wall of the orbit to provide sensory innervation to the lacrimal gland, then continues anteriorly to supply the skin of the upper lateral eyelid, as well as the scalp and conjunctiva.

The frontal nerve is the largest branch of the ophthalmic nerve. It travels anteriorly along the roof of the orbit, branching into the supraorbital and supratrochlear nerves, both of which supply skin of the upper medial eyelid, forehead, and mucosa of the frontal sinus.

The nasociliary nerve gives off several branches that supply the eye (ciliary nerves) and ethmoid and sphenoid sinuses as it courses medially and anteriorly. Reaching the anterior orbit, it gives off the infratrochlear nerve, supplying the medial skin of the eye lid, as well as the skin of the lateral aspect and tip of the nose.

The maxillary nerve (CN V2) contributes to sensory innervation around the eye through the infraorbital and zygomatic nerves. As the maxillary nerve approaches the orbit, it gives off the infraorbital nerve, which passes through the inferior orbital fissure, then travels along the infraorbital groove on the floor of the orbit, before passing through the infraorbital canal and infraorbital foramen to the skin. The infraorbital nerve provides sensory innervation to the skin of the cheek and lower eyelid.

The zygomatic nerve branches from the maxillary nerve and also enters the orbit through the inferior orbital fissure, where it then divides into the zygomaticofacial and zygomaticotemporal nerves. These nerves travel through the zygomatic foramen to provide sensory innervation to the lateral region of the eyelid.

Figure Description

Right orbit with overlying sensory nerves

Key

A zygomaticotemporal nerve
B zygomaticofacial nerve
C lacrimal nerve
D supraorbital nerve
E supratrochlear nerve
F infratrochlear nerve
G infraorbital nerve

Eyelid Innervation

sensory

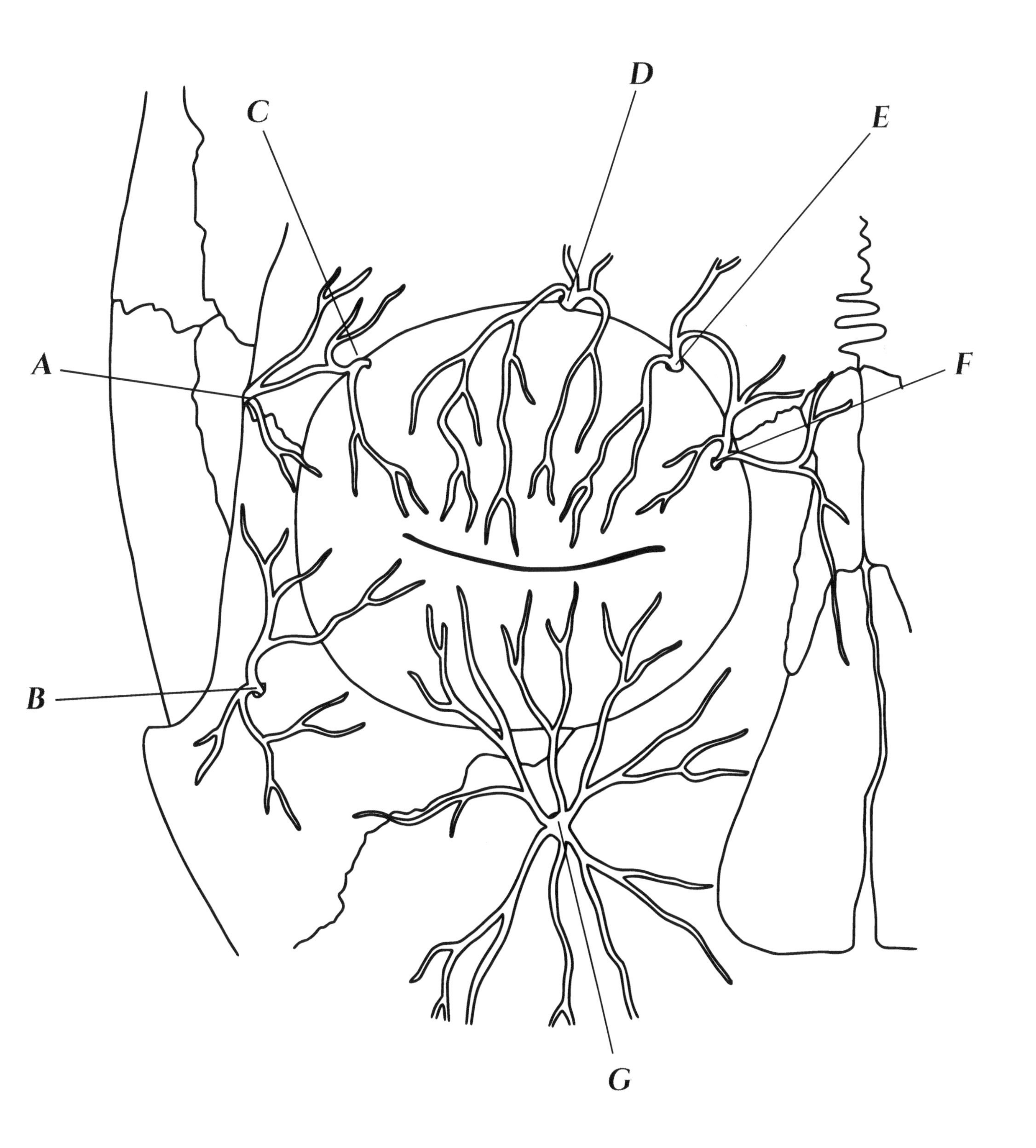

Tarsi of Eyelid

The dense irregular connective tissue of the orbital septum, the aponeurosis of the levator palpebrae superioris muscle, and tarsal plate make up the "skeleton" of the eyelid. The tarsal plate surrounds the tarsal (Meibomian) glands. In the upper eyelid, the tarsal plate is approximately 10 mm vertically at its center, merging with the aponeurosis of the levator muscle and the orbital septum at its superior aspect. The tarsal plate of the upper eyelid helps to maintain the concave shape that holds the eyelid taught against the globe. The tarsal plate of the lower lid is approximately 5 mm vertically at the center.

The medial and lateral ligaments serve to support and stabilize the lateral and medial margins of the tarsal plates. The medial palpebral ligament attaches to the maxillary bone, and the lateral palpebral ligament attaches to Whitnall's ligament. Whitnall's ligament is a transverse ligament in the orbital region of the upper eyelid that spans from the trochlea of the superior oblique muscle to the lateral wall of the bony orbit.

The orbital septum is found in the orbital region of the eyelid. The orbital septum serves as a barrier between the anterior layers of the eyelid and the posterior layers of the eyelid and orbit. The orbital septum prevents infection from spreading from the anterior eyelid to the orbit, and also helps to contain periorbital fat within the orbital regions of the eyelid.

Analogous to Whitnall's ligament in the upper eyelid is Lockwood's ligament in the lower eyelid (not shown). Lockwood's ligament is a transverse ligament that forms a support hammock for the eye. Anatomically, it is seen as a thickening of Tenon's capsule.

Figure Description

Right orbit with connective tissue of eyelids

Key

A superior tarsus
B lateral palpebral ligament
C inferior orbital septum
D superior orbital septum
E aponeurosis of levator muscle
F medial palpebral ligament
G inferior tarsus
H Whitnall's ligament

Tarsi of Eyelid

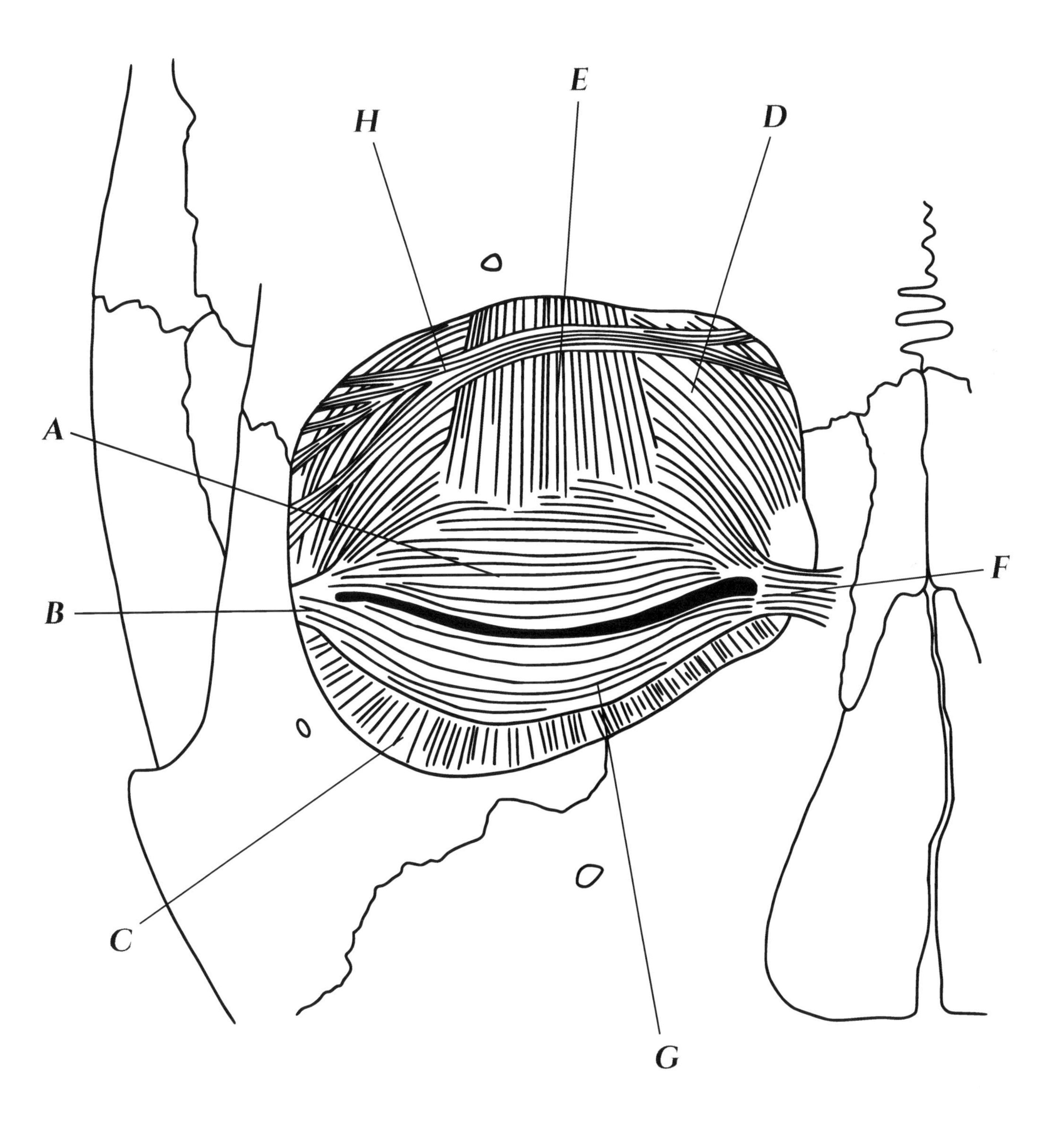

Meibomian Glands

The upper and lower eyelids contain vertically oriented tarsal glands, aka Meibomian glands, that secrete meibum at the lid margin. Meibum is a sebaceous product that contributes to the outermost layer of the tear film. The main components of meibum are lipids and proteins. Meibum helps to create a smooth refracting surface and slow evaporation of tears due to its hydrophobic properties.

The Meibomian glands are holocrine type exocrine glands, located in the more posterior aspect of the palpebral portion of the eyelids, contributing to the tarsal plate. The upper eyelid contains 30-40 glands, and the lower eyelid contains 20-30 glands.

The orifices of the Meibomian glands at the lid margin indicate the mucocutaneous junction, where the keratinization of the skin of the eyelid is lost. Anterior to the orifices is considered a "dry" surface, whereas posterior to the orifices is considered "wet."

Clinically, the Meibomian glands can be expressed by rolling a cotton swab along the length of the glands towards the lid margin. Healthy secretions are seen as clear and oily. In Meibomian gland dysfunction, secretions may appear thickened and opaque.

Figure Description

Isolated meibomian glands, adapted from Wolff, 1968

Key

A superior meibomian glands
B inferior meibomian glands
C palpebral fissure
D duct

Meibomian Glands

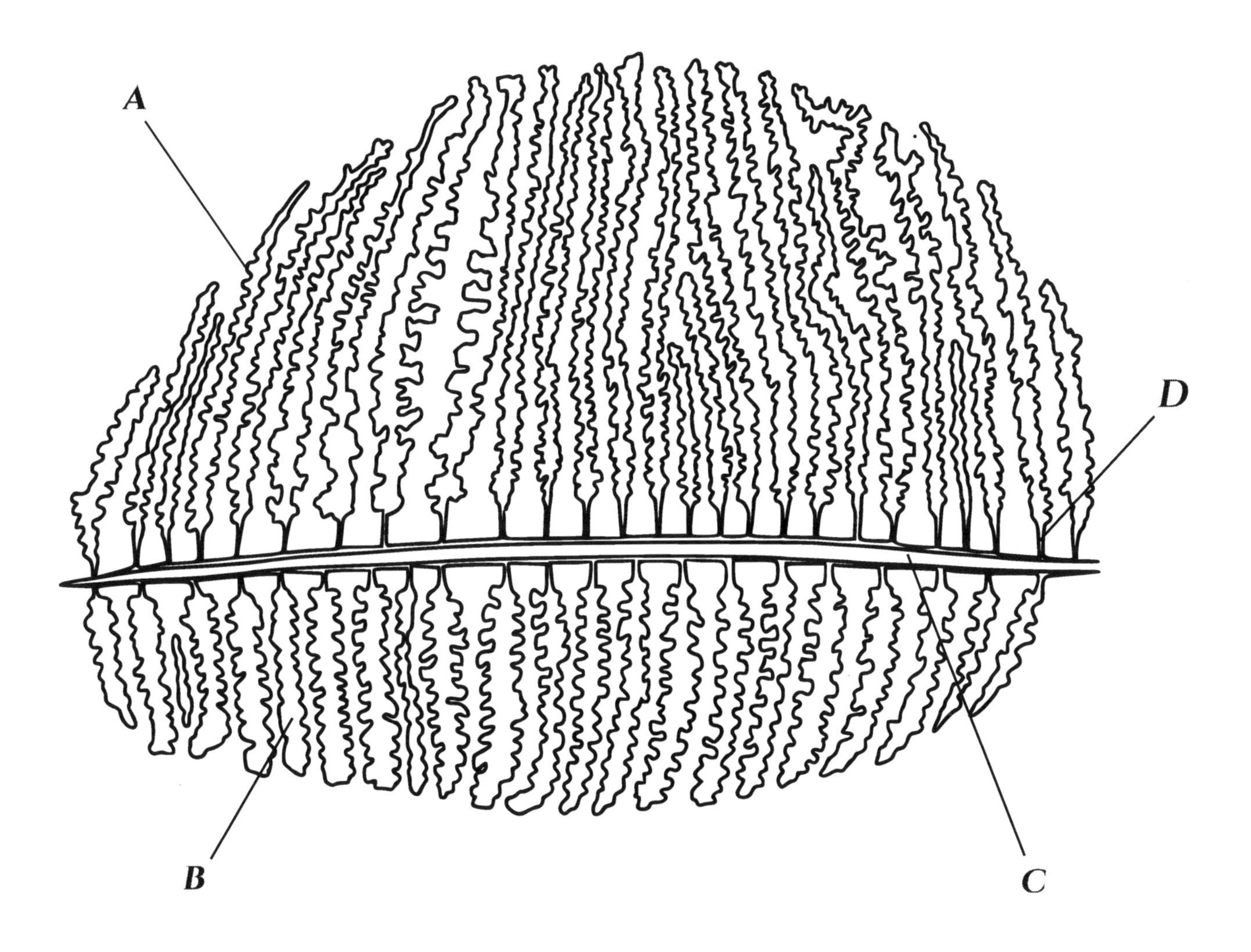

Eyelid Histology

The eyelids serve to protect the eye, functioning as a physical barrier blocking foreign bodies. The eyelids also limit light entering the eye and produce, spread, and drain tears. The eyelids span from the eyebrow superiorly to the upper cheek inferiorly. In cross section, the anterior most layer of the eyelid is a superficial keratinized skin layer. Posterior to the skin is a subcutaneous areolar layer, followed by a layer of muscles and connective tissue, a submuscular areolar layer, and, most posteriorly, the palpebral conjunctiva.

The eyelid is thin and elastic. It is richly vascularized and innervated. The vessels and nerves primarily travel through the areolar layers. Unique to the palpebral portion of the eyelid, the areolar layers do not contain adipose tissue.

There are three muscles of the eyelid, the orbicularis oculi, levator palpebrae superioris, and tarsal muscle, aka Müller's muscle. The orbicularis oculi is a striated muscle that closes the eyelids, innervated by the facial nerve (CN VII). The orbicularis oculi muscle encircles the palpebral fissure and spans both the orbital and palpebral portions of the eyelid. The orbital portion is responsible for forced blinking, and the palpebral portion functions in reflexive blinking. The ciliary portion of the muscle, aka muscle of Riolan, is found near the lid margin. The ciliary portion helps to keep the lid margin opposed to the globe.

The levator palpebrae superioris is a striated muscle that elevates the upper eyelid, innervated by the oculomotor nerve (CN III). The levator muscle is only found in the superior eyelid, not the inferior eyelid.

The tarsal muscle is a smooth muscle that retracts the lids, innervated by the sympathetic nervous system. The tarsal muscle is also known as Müller's muscle.

The orbital septum, aponeurosis of the levator muscle, and tarsal gland connective tissue contribute to the "skeleton" of the eyelid, dividing the lid into pre-septal and post-septal regions. This boundary is important for compartmentalizing the eyelid.

Figure Description

Cross section of a superior eyelid, palpebral region, adapted from Wolff, 1968

Key

A skin
B hair follicle
C sweat gland
D subcutaneous areolar
E eyelash follicle
F cilia (eyelash)
G gland of Moll
H gland of Zeis
I orbicularis oculi, orbital region
J orbicularis oculi, palpebral region
K orbicularis oculi, ciliary region, aka muscle of Riolan
L preaponeurotic adipose tissue
M orbital septum
N aponeurosis of the levator palpebrae superioris muscle
O tarsal muscle, aka Müller's muscle
P gland of Krause
Q peripheral arterial arcade
R gland of Wolfring
S tarsal gland, aka Meibomian gland
T marginal arterial arcade
U gray line
V tarsal gland orifice
W mucocutaneous junction
X lid wiper region
Y subtarsal fold
Z palpebral conjunctiva
AA goblet cells
AB Stieda's grooves
AC fornix conjunctiva

The lid margin is approximately 2 mm wide. Two to three rows of cilia (eyelashes) can be found most anteriorly. Posterior to the cilia is the gray line, followed by the Meibomian gland orifices. Rounding the posterior edge of the lid margin are the lid wiper region and subtarsal fold. The lid wiper region is held taught to the ocular surface via the ciliary portion of the orbicularis oculi muscle. The subtarsal fold designates the complete transition of the keratinized stratified squamous epithelium of the skin to the mucous membrane of the palpebral conjunctiva.

Eyelid Histology

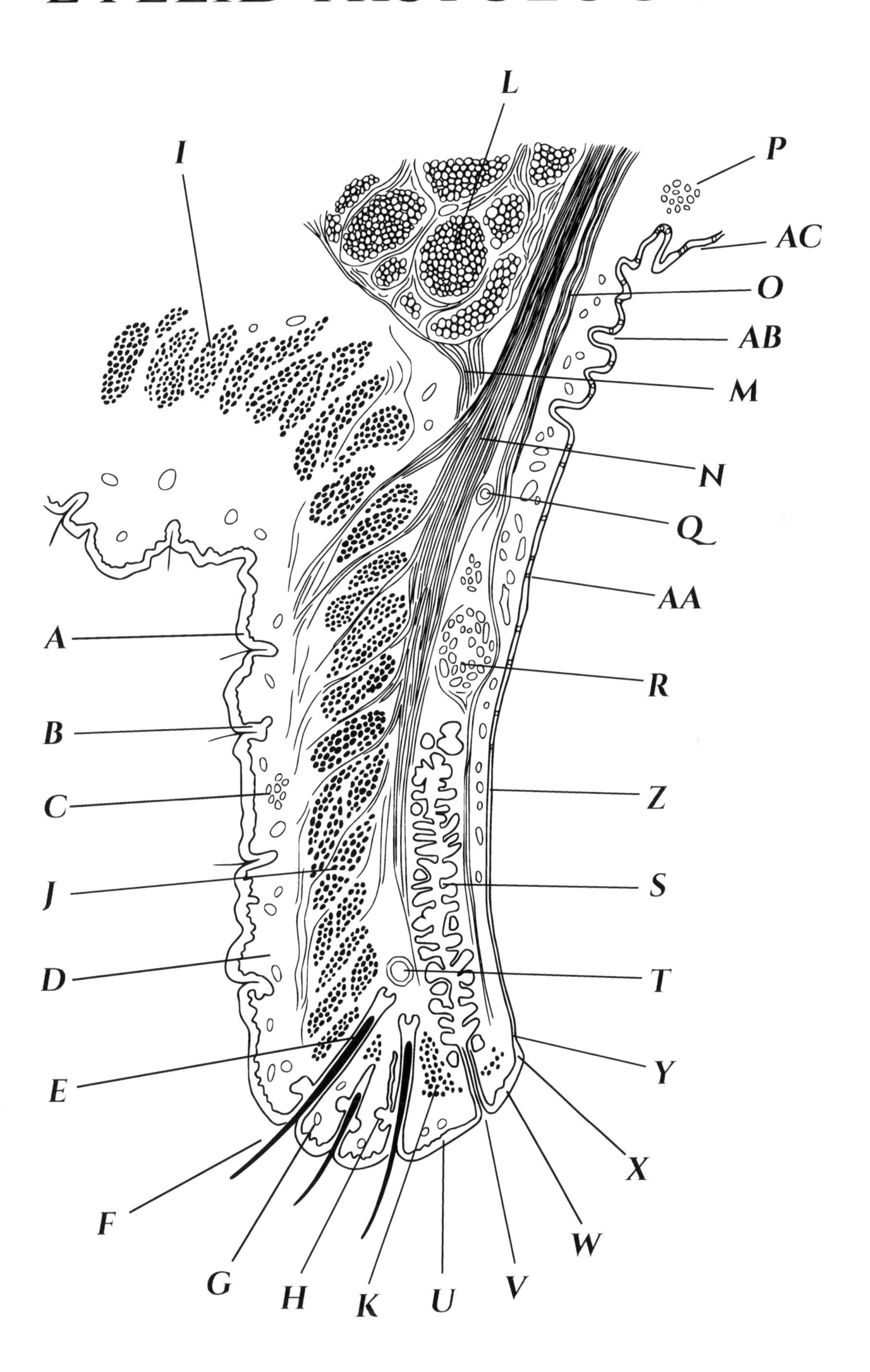

The Anterior Segment

Conjunctiva

The conjunctiva is a thin transparent mucous membrane that lines the posterior eyelid and anterior eye. The conjunctiva is composed of a non-keratinized stratified epithelium and stroma. The conjunctiva is the most immunologically active tissue of the eye and is richly vasculaturized, making the conjunctiva susceptible to hyperemia and inflammation with irritation, allergy, or injury.

The conjunctiva is divided into three regions, palpebral, bulbar, and fornix. The palpebral conjunctiva lines the posterior surface of the eyelids. The bulbar conjunctiva lines the anterior surface of the sclera. The fornix conjunctiva is the transition between the bulbar and palpebral conjunctiva.

The palpebral conjunctiva is continuous with the skin of the eyelids. Posterior to the tarsal gland orifices on the lid margin, the skin loses its keratinization and begins to transition into conjunctiva as the epithelium courses around the posterior aspect of the lid margin. The palpebral conjunctiva consists of 2-3 layers of stratified squamous and cuboidal epithelium.

The epithelium of the fornix conjunctiva is more columnar. The fornix conjunctiva forms a pocket, known as the conjunctival sac. The depth of the conjunctival sac is greatest in the superior fornix at 14-15 mm. The inferior fornix is 10-12 mm deep, and the lateral fornix is 5-8 mm deep. The medial fornix has no depth due to the structures found at the medial canthus, including the caruncle and plica semilunaris.

The bulbar conjunctiva lines the anterior aspect of the eye, laying over Tenon's capsule and the sclera. The epithelium of the bulbar conjunctiva is more squamous and is continuous with the epithelium of the cornea.

Mucin-secreting goblet cells are found within the epithelium of the conjunctiva. The goblet cell density is greatest in the nasal conjunctiva and fornices.

Figure Description

Left eye, inferior lid depressed to reveal inferior fornix

Key

A nasal bulbar conjunctiva
B inferior palpebral conjunctiva
C inferior fornix
D temporal bulbar conjunctiva
E inferior lid margin
F superior lid margin
G caruncle
H plica semilunaris

CONJUNCTIVA

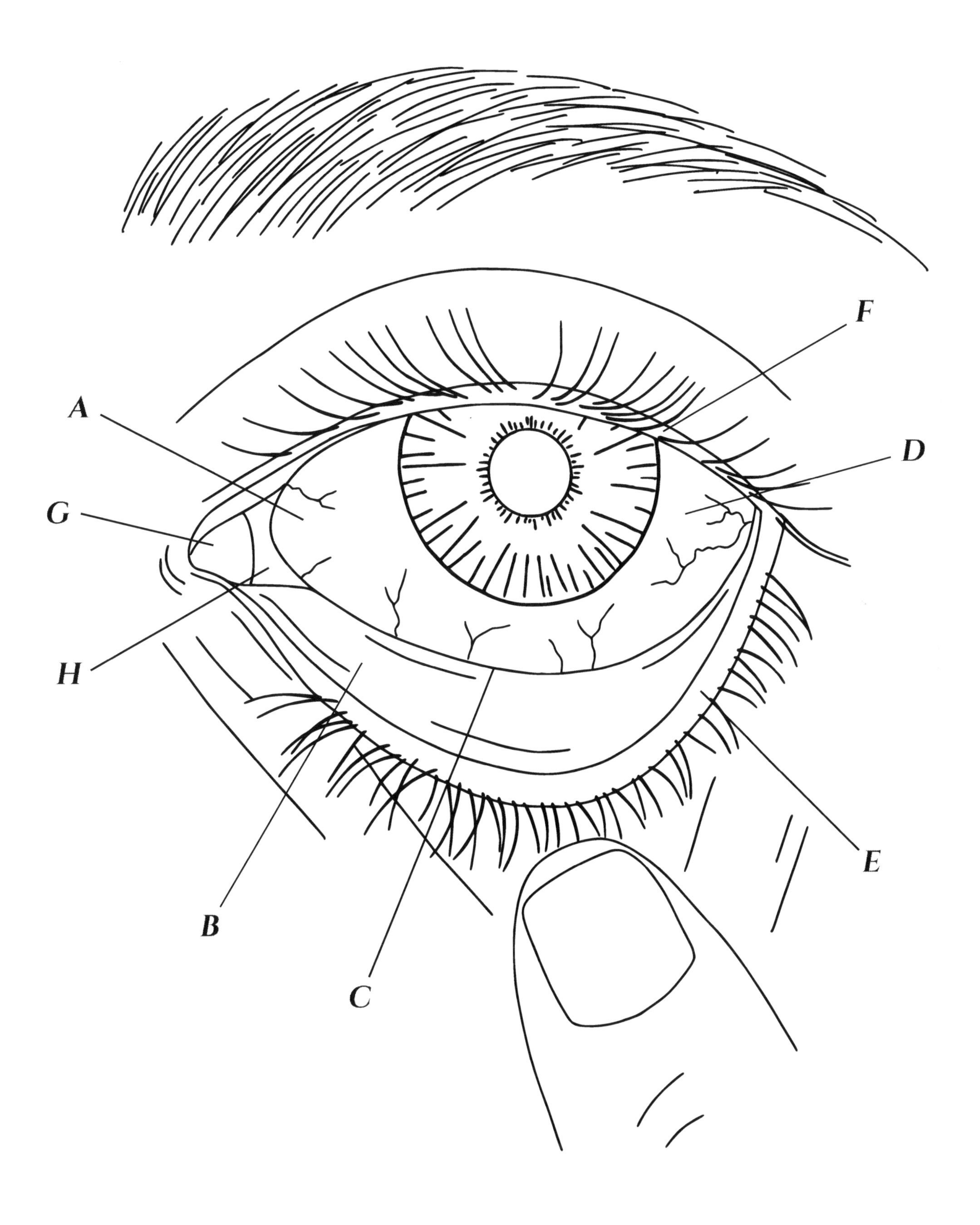

Conjunctiva Histology

The conjunctiva consists of epithelial and stromal layers. The epithelium is non-keratinized and stratified, with 2-4 layers of cells. The epithelium is squamous/cuboidal in the palpebral and bulbar regions and columnar in the fornices. The epithelium contains a significant number of goblet cells, neutrophils, and lymphocytes. The superficial epithelial cells contain microvilli and may have melanin granules.

Goblet cells are unicellular mucin-secreting holocrine glands that, in the eye, primarily secrete MUC5AC, contributing to the innermost layer of the tear film. Goblet cells are rounded, with microvilli on the apical portion where the secretory granules are localized. The cell organelles are localized towards the basal aspect of the cell.

The stroma, or submucosa, of the conjunctiva consists of loose connective tissue with an outer lymphoid (adenoid) layer and inner fibrous layer. The outer lymphoid layer contains numerous capillaries, as well as immunological structures, including immunoglobulin A, lymphocytes, mast cells, leukocytes, and eosinophils. The deep fibrous layer has a denser collagen and fibroblast content, with blood vessels, nerves, and lymphatic vessels. Lymphatics of the conjunctiva drain to the superficial preauricular and deep submaxillary lymph nodes. The accessory lacrimal glands of Wolfring and Krause are found in the fibrous layer of the palpebral and fornix conjunctiva, respectively.

Blood supply to the palpebral conjunctiva is provided by the dorsonasal and palpebral arteries, which form arcades within the eyelids. Blood supply to the bulbar conjunctiva is provided by the anterior ciliary arteries. Venous drainage occurs via the anterior ciliary veins and superior and inferior ophthalmic veins.

The conjunctiva contains sensory and autonomic innervation. Sensory innervation to the palpebral and fornix conjunctiva is provided by the lacrimal, supraorbital, supratrochlear, and infratrochlear nerves, which are branches of the ophthalmic nerve (CN V1), and the infraorbital nerve, which is a branch of the maxillary nerve (CN V2). Sensory innervation to the bulbar conjunctiva is provided by the ciliary nerves. Sympathetic innervation to the blood vessels is via the superior cervical ganglion, and parasympathetic innervation is via branches of the facial nerve (CN VII), with post-ganglionic fibers from the pterygopalatine ganglion.

Figure Description

Histology of the conjunctiva

Key

A epithelium
B adenoid layer of stroma
C fibrous layer of stroma
D goblet cell
E mast cell
F lymphocytes
G vasculature
H nerve

Conjunctiva Histology

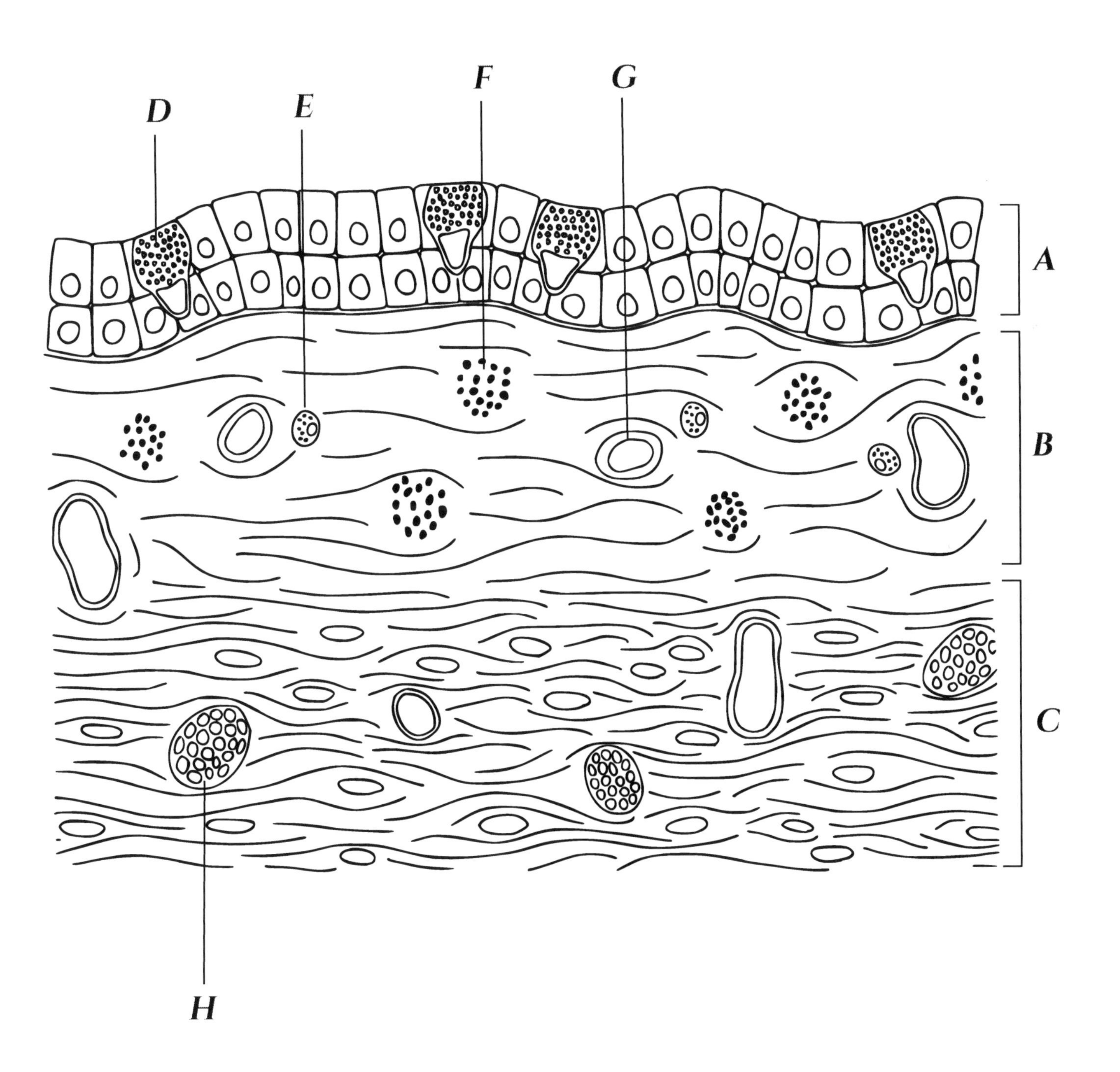

The Cornea

The cornea is the avascular transparent anterior-most structure of the eye that functions to refract light and protect the eye. The cornea covers about 1/6 of the surface area of the eye. The average anterior surface horizontal diameter is 11.7 mm, and the vertical diameter is 10.6 mm. The posterior corneal surface has a horizontal and vertical diameter of 11.7 mm.

The curvature of the anterior surface is steeper than that of the posterior surface. Because of this geometry, the cornea is thinnest at the center, with an average central corneal thickness of 550 μm, and thickest at the periphery, with an average thickness of 650 μm. The average radius of curvature of the anterior surface is 7.8 mm, and of the posterior surface is 6.4 mm. The refractive power of the cornea is 40-44 diopters. Thus, the cornea is responsible for about 2/3 of the total refracting power of the eye. Often, the cornea is not completely spherical, but rather, toric in nature, giving rise to an astigmatic refractive error. More commonly, the cornea is steeper in the vertical meridian, leading to with-the-rule astigmatism. Against-the-rule astigmatism occurs when the cornea is steeper in the horizontal meridian.

The cornea meets the conjunctiva and sclera at the limbus. Deep to the limbus is the iris root. The posterior cornea and iris root form the angle of the eye, known as the iridocorneal angle, where aqueous humor drainage structures are located.

The refractive index of the cornea is 1.376. The refractive index of air is 1.0, and of the tear film is 1.336. Therefore, the main refracting surface is the tears, as light passes from air to tears. The cornea transmits all visible wavelengths, 400-700 nm, as well as most ultraviolet A, 315-400 nm, and some ultraviolet B, 280-315 nm. About 50% of light at 310 nm is transmitted by the cornea, while no light less than 280 nm is transmitted.

The cornea is a relatively dehydrated structure, consisting of 78% water, 15% collagen, and 5% proteins.

Figure Description

Cross section of the anterior segment

Key

A central anterior corneal surface
B central posterior corneal surface
C peripheral cornea
D anterior chamber
E iridocorneal angle
F iris root

THE CORNEA

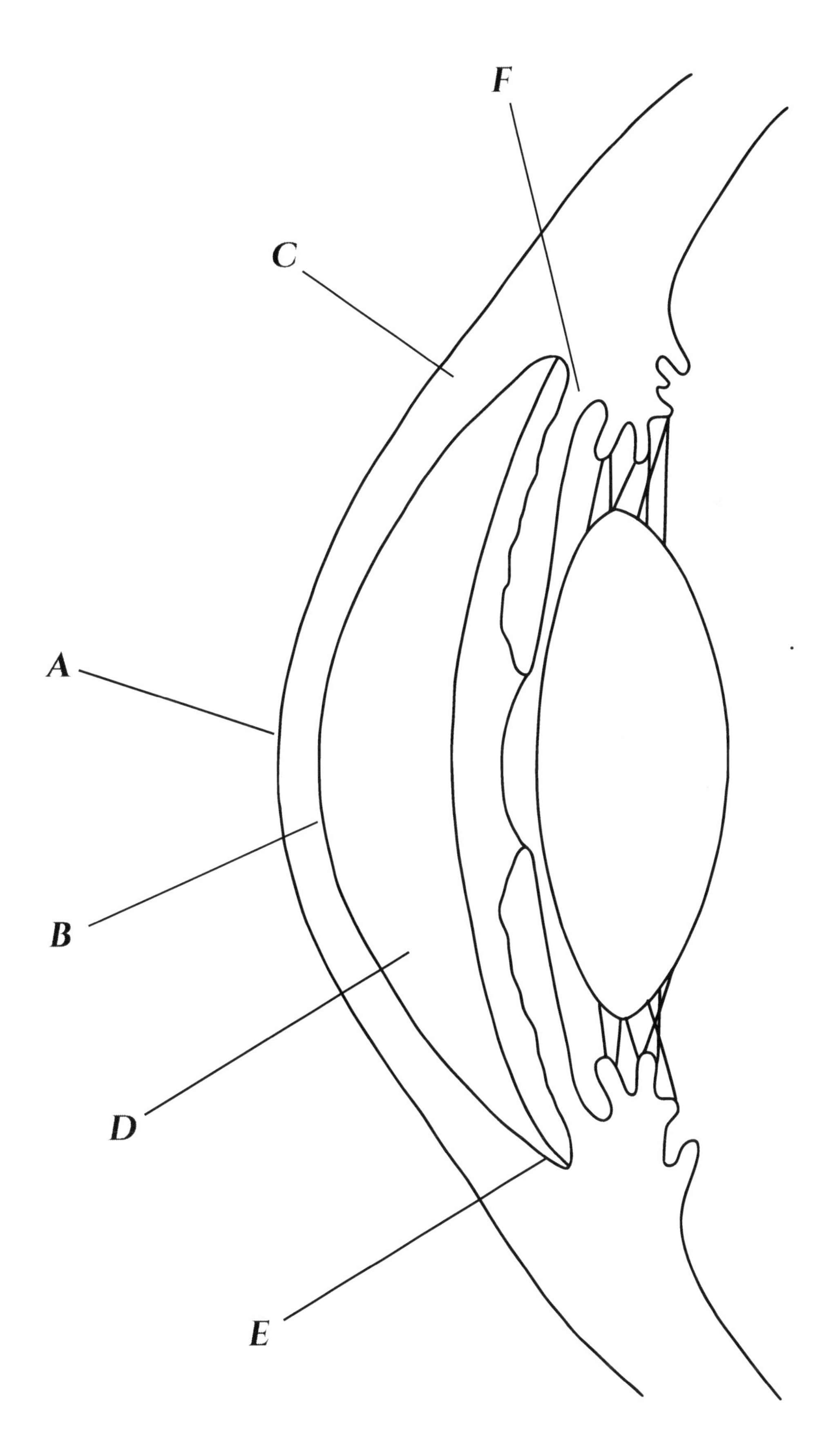

Cornea Layers

The cornea consists of 5 layers. From anterior to posterior, the layers are epithelium, anterior limiting lamina (aka Bowman's membrane), stroma, posterior limiting lamina (aka Descemet's membrane), and endothelium.

The epithelium consists of 5-7 layers of stratified cells and a basement membrane. It is approximately 50 μm thick, making up about 10% of the thickness of the cornea.

The anterior limiting lamina is posterior to the epithelium and is 8-14 μm thick. The anterior limiting lamina is acellular and consists of dense type I collagen.

The stroma is approximately 450 μm thick, making up about 90% of the thickness of the cornea. The stroma consists of type I collagen, keratocytes, ground substance, and water.

The posterior limiting lamina is the basement membrane of the endothelium. It is composed of type IV collagen and grows throughout life. At birth, it is about 3 μm thick, and can reach up to 15 μm later in life.

The endothelium is a single layer of squamous cells spanning the posterior surface of the cornea.

Figure Description

Histology of the cornea, full thickness

Key

- A epithelium
- B anterior limiting lamina (Bowman's membrane)
- C stroma
- D nucleus of keratocyte
- E posterior limiting lamina (Descemet's membrane)
- F endothelium

Corneal Layers

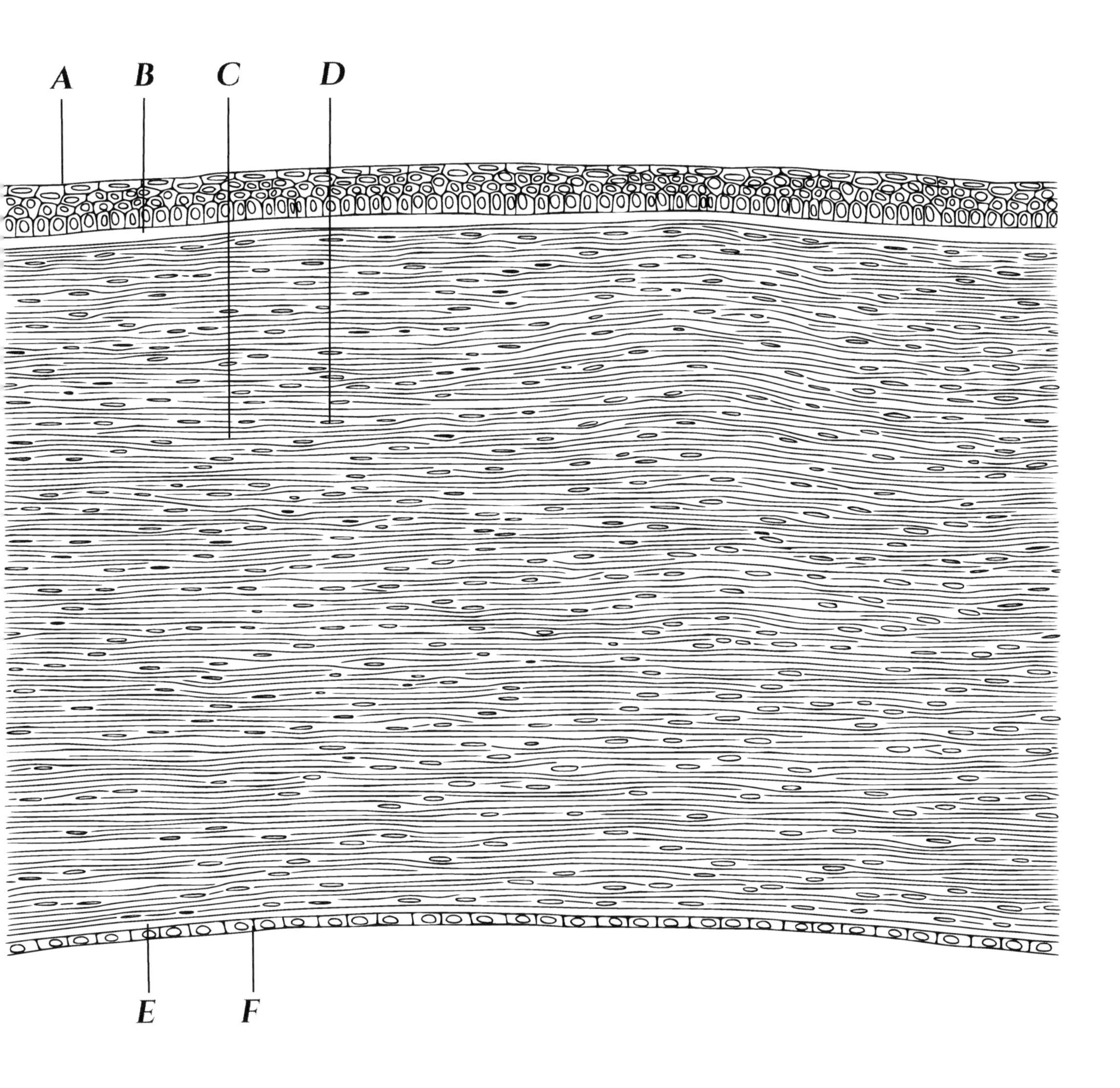

Corneal Epithelium and Anterior Limiting Lamina

The corneal epithelium forms a barrier between the environment and the cornea. It is a non-keratinized stratified epithelium of 5-7 layers of cells, with a total thickness of approximately 50 μm. The corneal epithelium is derived from surface ectoderm in development.

The posterior most epithelial cells are columnar in shape and sit on a basement membrane. Given their shape and location, they are called basal cells. They have a flat basal surface and rounded head. Moving anteriorly, the cells become progressively flatter. The middle epithelial cells are called wing cells, due to their more polyhedral shape and processes that meet neighboring cells. The anterior cells are called squamous cells due to their flattened shape. The most anterior squamous cells have microvilli and microplicae on the apical surface with a glycocalyx matrix that helps hold the inner mucin layer of the tear film to the corneal surface.

Corneal epithelial cells are labile; they migrate and change shape. The basal layer of cells is continuously undergoing mitosis, then migrating anteriorly. As the cells migrate, they transition into wing cells and squamous cells, eventually sloughing off from the corneal surface in the process of desquamation. Approximately 4% of the cells are undergoing mitosis in a healthy cornea, such that corneal cells have a 7-10 day turnover.

Through the process of migration, the cells maintain their barrier function through numerous cell to cell junctions. These junctions allow the cells to be mechanically strong, resisting shearing forces such as blinking and rubbing the eyes. The cell walls have interdigitations that help maintain connections with neighboring cells. The basal cells are adhered to the basement membrane through hemidesmosomes. Desmosomes and gap junctions are found throughout the cornea. Squamous cells are joined to each other with zonula occludens (tight junctions) to form a tight barrier, making the surface impermeable to small molecules and fluid.

The corneal epithelium is the most densely innervated structure of the body. Sensory innervation is carried by the long ciliary nerves to the nasociliary nerve, a branch of the ophthalmic division of the trigeminal nerve (CN V).

Figure Description

Histological section of corneal epithelium, adapted from Hogan, Alvarado, Weddell, 1971

Key

- A superficial squamous cells
- B wing cells
- C columnar basal cells
- D anterior limiting lamina (Bowman's membrane)
- E desmosome
- F tight junction
- G microvilli
- H gap junction
- I nerve
- J hemidesmosome
- K basement membrane
- L lymphocyte

Nerves lose their myelination as the enter the corneal stroma, and further lose their Schwann cell covering as they approach the epithelium, traveling in the epithelium as bare nerve endings.

The anterior limiting lamina, also known as Bowman's layer, sits just posterior to the epithelial basement membrane. The anterior limiting lamina is an acellular transition layer between the epithelium and stroma. It is 8-14 μm and consists of dense type I collagen. The collagen fibrils are smaller in diameter than those in the stroma and more irregularly organized. This layer is resistant to mechanical forces; however, it does not renew. Therefore, if this layer is damaged, a scar may form. The anterior limiting lamina is not necessary for healthy corneal function. The anterior limiting lamina is removed during photorefractive keratectomy, a laser refractive surgery, with no consequence to the health of the cornea.

Corneal Epithelium and Anterior Limiting Lamina

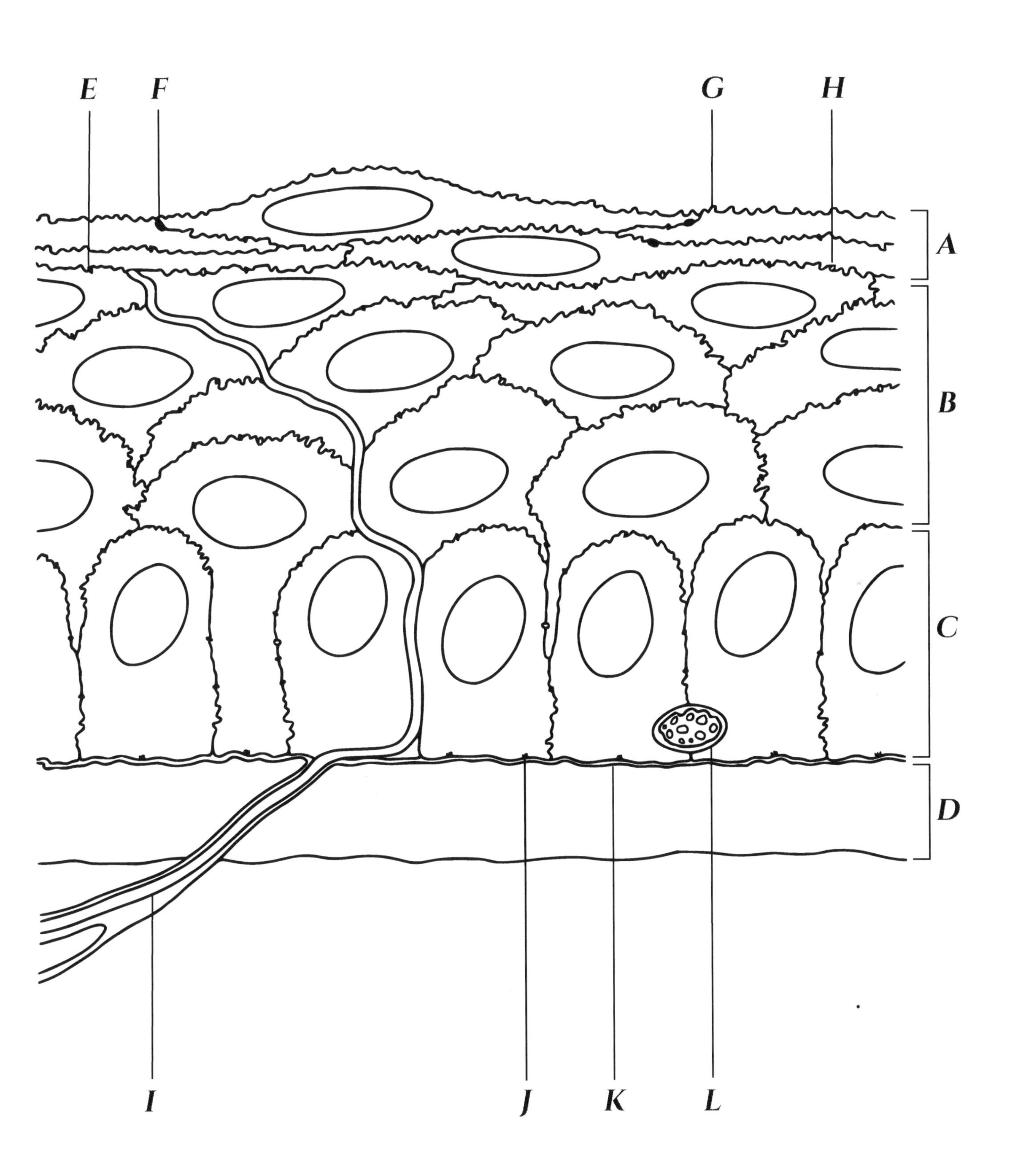

Corneal Stroma

The corneal stroma (substantia propria) is the thickest layer of the cornea, at about 450 μm, comprising 90% of the corneal thickness. The stroma consists of regularly organized layers of, primarily, type I collagen fibrils. Keratocytes and ground substance are interspersed with the collagen fibrils. The corneal stroma is continuous with the sclera; however, the collagen of the sclera is more dense and irregularly organized. The corneal stroma is derived from neural crest in development.

In the corneal stroma, collagen fibrils are about 30 nm in diameter and are organized in lamellae, or layers, of about 1 μm each. The collagen fibrils of the anterior 1/3 of the stroma are more interwoven and have more crosslinking than the posterior 2/3. The collagen lamellae of the stroma run parallel to the corneal surface and span from limbus to limbus in a criss-cross pattern, such that subsequent lamellae run in a perpendicular direction to the previous lamellae. There are approximately 242 lamellae in the human cornea. Spacing of collagen fibrils is maintained in part by a ground substance that contains of glycosaminoglycans.

The corneal stroma tends to attract water, but must maintain a state of deturgescence, or dehydration, for transparency. The stroma should maintain a water content of approximately 78% for transparency. The 1957 transparency theory by Maurice says that to maintain transparency, the collagen fibrils must be arranged in a lattice/hexagonal pattern with a constant separation. Goldman and Benedek modified this theory in 1967 to say that while the spacing between collagen fibrils does not need to be exactly uniform, it must be less than half the wavelength of light.

Figure Description

Histology of the corneal stroma, adapted from Hogan, Alvarado, Weddell, 1971

Key

A collagen fibrils running longitudinally
B collagen fibrils in cross section

Corneal Stroma

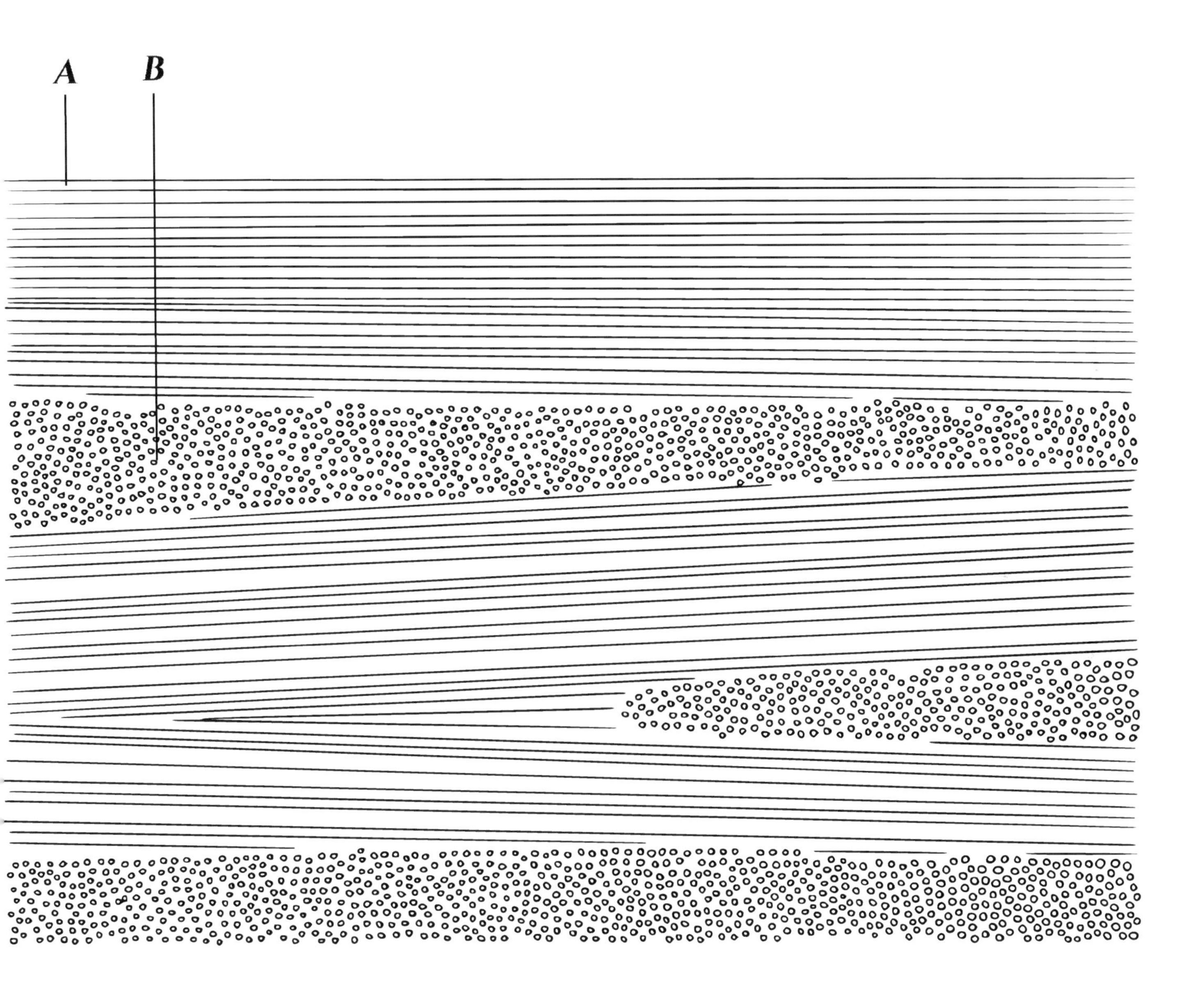

Keratocytes

Keratocytes serve the function of fibroblasts in the corneal stroma. Keratocytes are involved in matrix turnover, intracorneal communication, interlamellar tethering, wound healing, and phagocytosis, and they serve as a glycogen reservoir.

There are approximately 2.4 million keratocytes in each cornea, making up about 10% of the stroma. The keratocytes are located between the collagen lamellae. They serve to maintain the stroma by secreting collagen fibrils and an extracellular matrix of proteoglycans, which are macromolecules of glycosaminoglycans.

The keratocytes are more dense in the anterior stroma. In an en face view, the keratocytes are arranged in a corkscrew pattern. The cell bodies are parallel to the corneal surface, and the stellate processes reach out and contact neighboring keratocytes via gap junctions.

Figure Description

Keratocytes

Key

A	keratocyte
B	nucleus of keratocyte
C	gap junction

KERATOCYTES

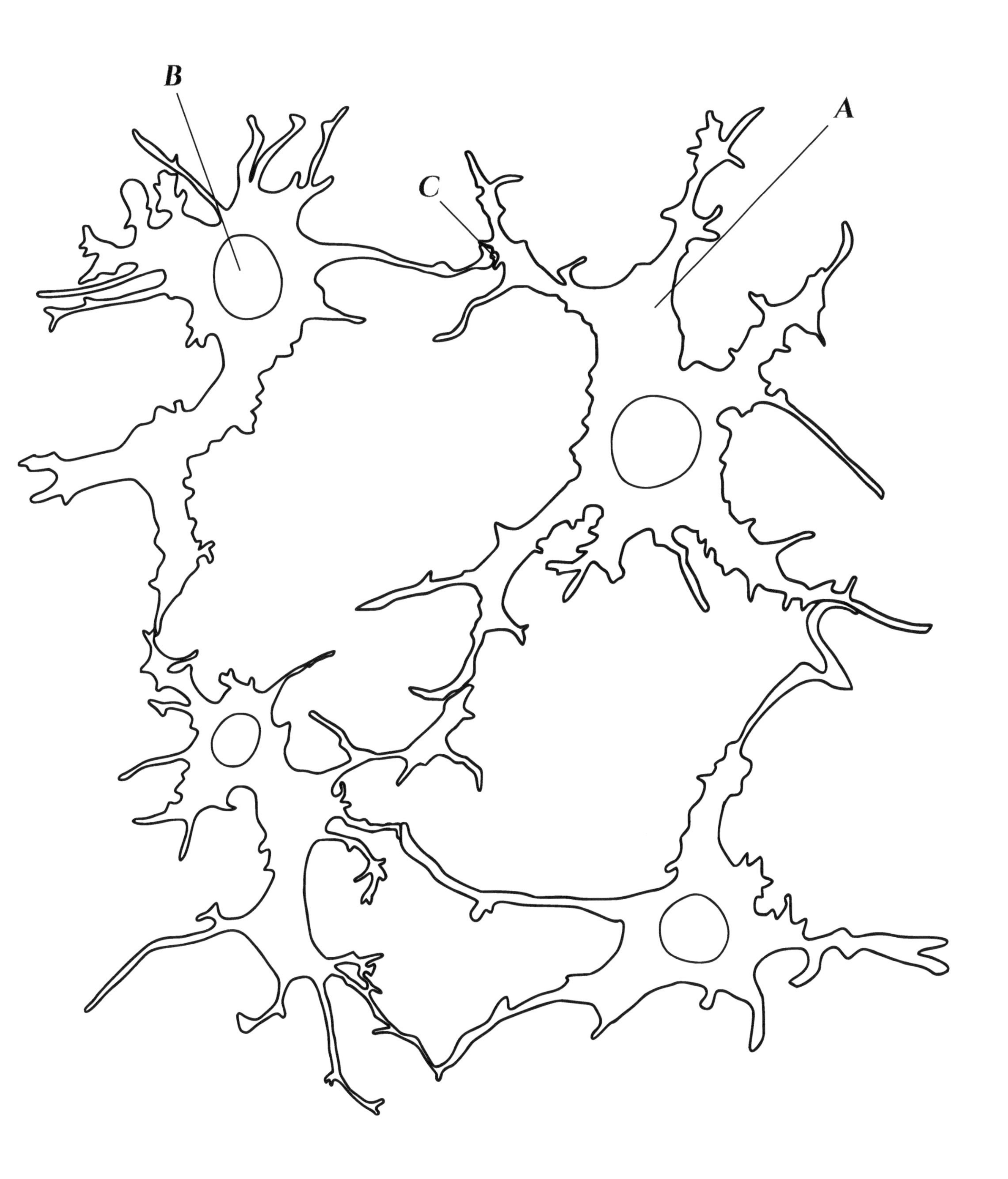

Corneal Endothelium

The corneal endothelium is a single layer of hexagonal cells spanning from limbus to limbus, providing complete coverage of the posterior aspect of the cornea. The apical surface of the endothelial cells faces the anterior chamber. The corneal endothelium is derived from neural crest in development.

Corneal endothelial cells perform an important role in maintaining deturgescence of the cornea through the essential pump mechanism. The cells prevent an influx of salts into the stroma to minimize water absorption by osmosis. Osmotic pressure is actively reduced through sodium potassium pumps (3 sodium ions out for 2 potassium ions in) and bicarbonate channels.

Endothelial cells cannot reproduce and are lost throughout life. As cells are lost, the remaining cells spread out to provide coverage. There are approximately 4200 cells/mm^2 at birth, which decreases at a rate of about 0.6% per year, stabilizing around 3000 cells/mm^2 in adulthood. Neighboring cells have tight, intermediate, and gap junctions.

With hypoxia, endothelial cells die and the remaining cells spread out, leading to polymegathism (various sized cells) and pleomorphism (various shaped cells). The functional limit is 700 -1000 cells/mm^2.

Figure Description

Corneal endothelium, en face

Corneal Endothelium

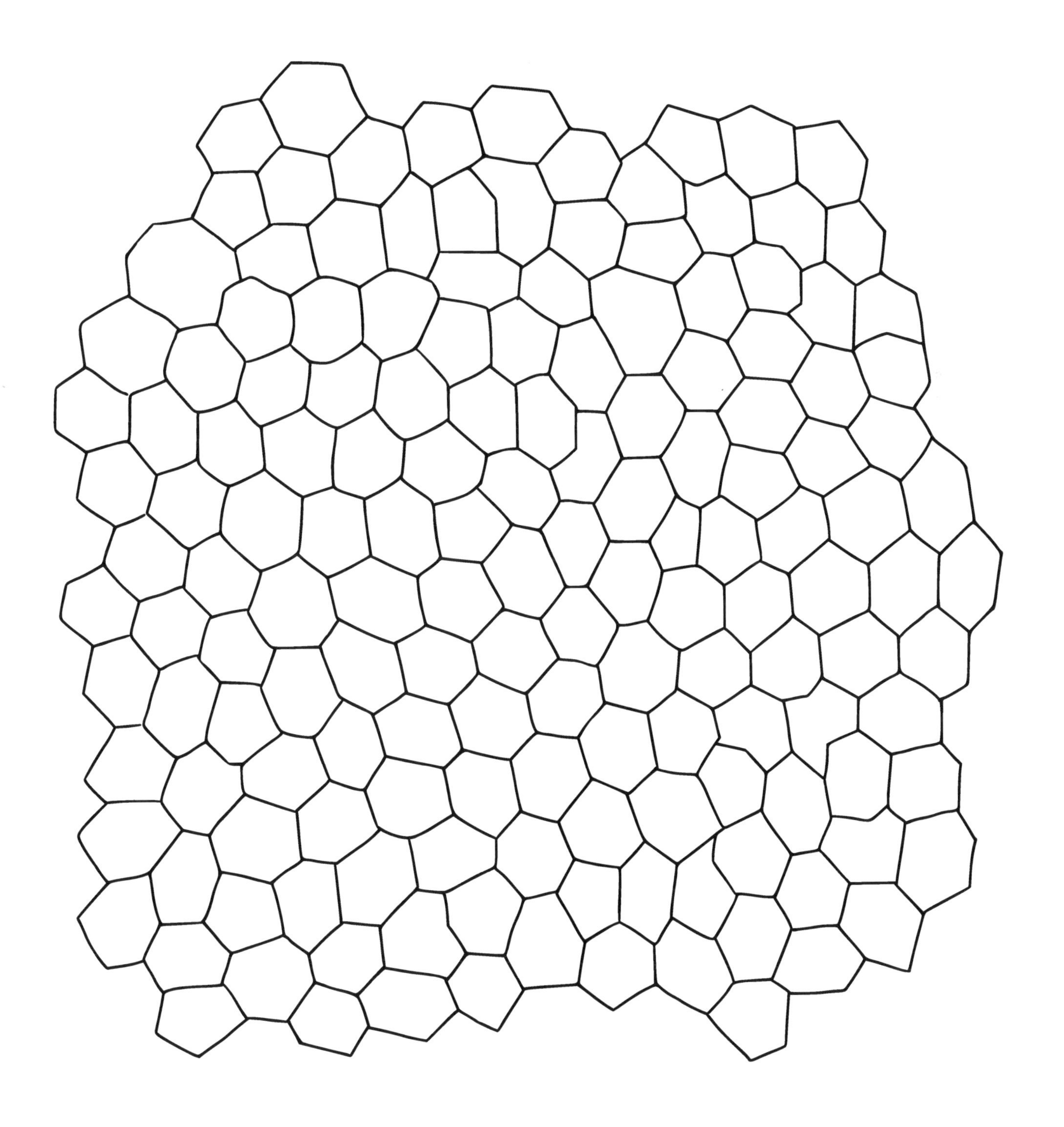

The Limbus

The limbus is a 1-2 mm transition zone between the transparent cornea and the opaque sclera. The anterior limit of the limbus, or corneolimbal junction, is demarcated by a "line" through the termination of the anterior and posterior limiting laminae of the cornea. The posterior limit of the limbus, or limboscleral junction, is demarcated by a "line" from the epithelial surface to the posterior aspect of the scleral spur. Deep to the limbus, the aqueous outflow structures are located in the iridocorneal angle, i.e. where the iris root meets the posterior peripheral cornea.

As the cornea transitions into the limbus, the epithelium thickens to up to 10 cell layers and becomes more undulated. Pigment is often found in the limbal epithelium. The undulations form radially oriented rete pegs with stromal channels in between, known as palisades of Vogt. Limbal stem cells reside in the palisades. Limbal stem cells divide and differentiate into corneal epithelial cells. The rate of division substantially increases during corneal wound healing. At the peripheral boundary of the limbus, the epithelium transitions into conjunctival epithelium. The conjunctival stroma, Tenon's capsule, episclera, and sclera are located deep to the conjunctival epithelium.

The corneal stroma continues into the limbus and sclera. In the cornea, the stroma is highly organized with small diameter collagen fibrils. Moving into the limbus, the collagen becomes less organized and more irregular, with larger fibrils. These morphological changes of the stroma contribute to the opacity of the sclera.

The vasculature of the limbus forms the limbal loops (aka marginal arcades) that supply the cornea and conjunctiva. The limbal plexus resides just deep to the limbal epithelium. More peripherally, vasculature includes a conjunctival plexus, and superficial and deep episcleral plexi. The limbal vasculature is primarily provided by the episcleral arterioles, which are branches of the anterior ciliary arteries.

Structures involved in aqueous outflow are found at the deepest portion of the limbus facing the anterior chamber. The aqueous outflow structures include the trabecular meshwork and Schlemm's canal.

Figure Description

Cross section of limbus, adapted from Hogan, Alvarado, Weddell, 1971

Key

A corneolimbal junction
B limboscleral junction
C limbal plexus
D conjunctival epithelium
E conjunctiva
F Tenon's capsule
G episclera
H sclera
I ciliary muscle
J Schlemm's canal
K trabecular meshwork
L scleral spur
M intramuscular arterial circle
N iris root
O major arterial circle

The Limbus

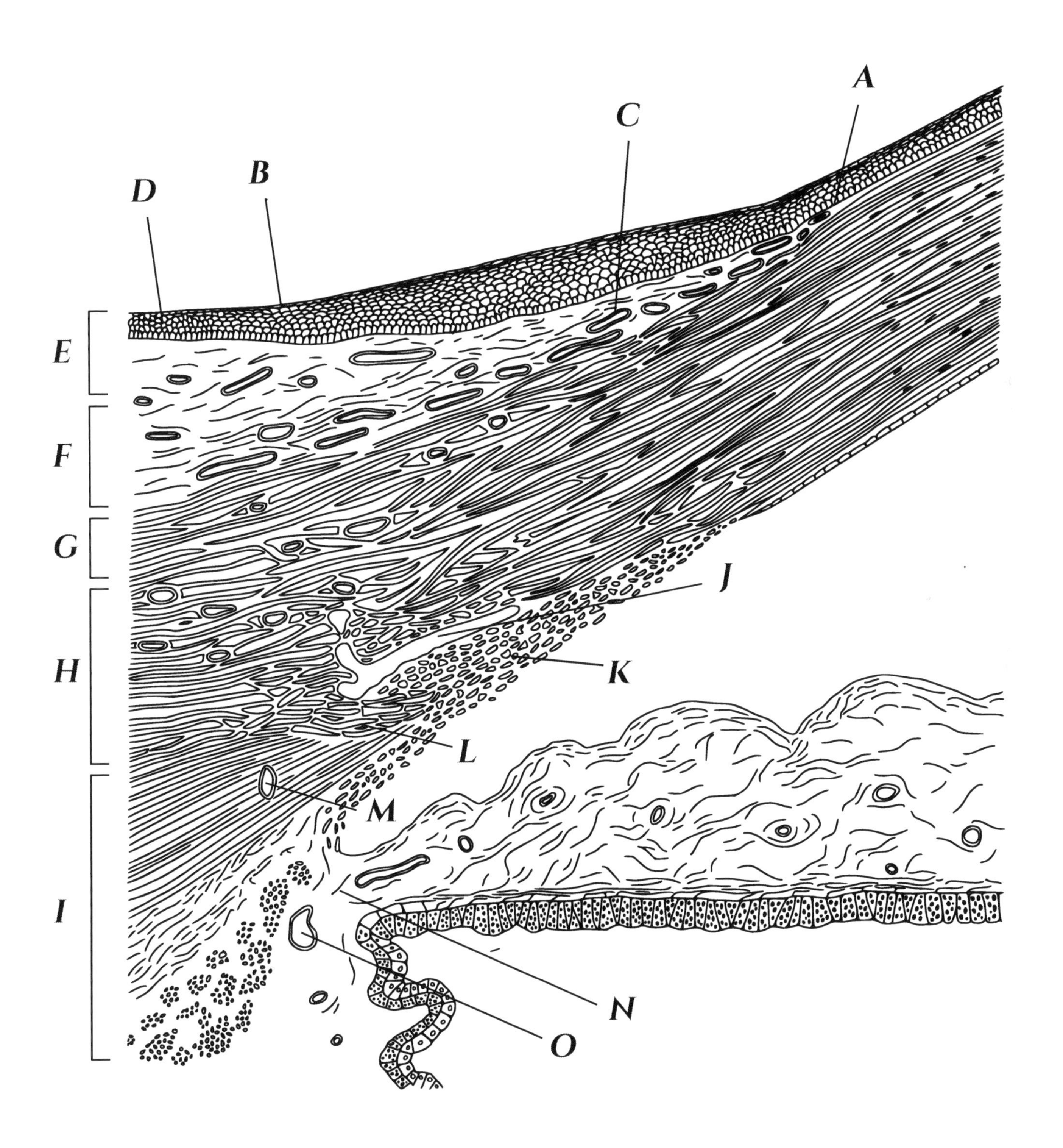

The Angle

The iridocorneal angle, aka internal scleral sulcus, is the junction of the peripheral cornea and iris root. The angle faces the anterior chamber and can be viewed clinically using a gonioscopy lens. All of the structures within the angle form a ring encircling the cornea, and can be seen in the following order: 1) Schwalbe's line, 2) trabecular meshwork, 3) scleral spur, 4) anterior face of the ciliary body, and 5) anterior surface of the iris. Schlemm's canal is generally not visible in a gonioscopic view.

The most anterior structure of the angle is Schwalbe's ring, which is the termination of the posterior limiting lamina of the cornea.

The next structure is the trabecular meshwork, which is the major site of aqueous humor filtration from the anterior chamber. The trabecular meshwork is a circumferential structure that is more narrow anteriorly, such that it appears triangular in cross section with the base of the triangle opposing the scleral spur and anterior face of the ciliary muscle. The external portion of the trabecular meshwork opposes Schlemm's canal and the interior portion faces the anterior chamber.

The trabecular meshwork has three regions, the uveal cords, the corneoscleral meshwork, and the juxtacanalicular meshwork. The uveal cords are adjacent to the anterior chamber and consist of cords of collagen, elastin, and ground substance surrounded by endothelial cells. The corneoscleral meshwork makes up the largest portion of the trabecular meshwork, spanning from the cornea to the scleral spur closer to Schlemm's canal. The pores are smaller in this region than within the uveal cords. The smallest pores are in the juxtacanalicular meshwork. Aqueous humor moves through the trabecular meshwork and into Schlemm's canal via a pressure gradient. In the juxtacanalicular region, the endothelial cells form vacuoles that carry the aqueous humor into Schlemm's canal.

The scleral spur is seen as a whitish region beneath the trabecular meshwork. The scleral spur is a circumferential band of scleral collagen that projects under Schlemm's canal. The scleral spur is the site of attachment for the longitudinal muscle fibers of the ciliary muscle.

The next structure in the angle is the anterior face of the ciliary body. This is seen as a highly pigmented band of tissue.

Figure Description

View into the iridocorneal angle, adapted from Hogan, Alvarado, Weddell, 1971

Key

A	Schwalbe's line/ring
B	trabecular meshwork
C	iris
D	cornea
E	posterior limiting lamina
F	corneal endothelium
G	limbus
H	external collector channel
I	Schlemm's canal
J	scleral spur (internal angle view)
J'	scleral spur (cross section)
K	longitudinal fibers of ciliary muscle
L	radial fibers of ciliary muscle
M	circular fibers of ciliary muscle
N	major arterial circle

The Angle

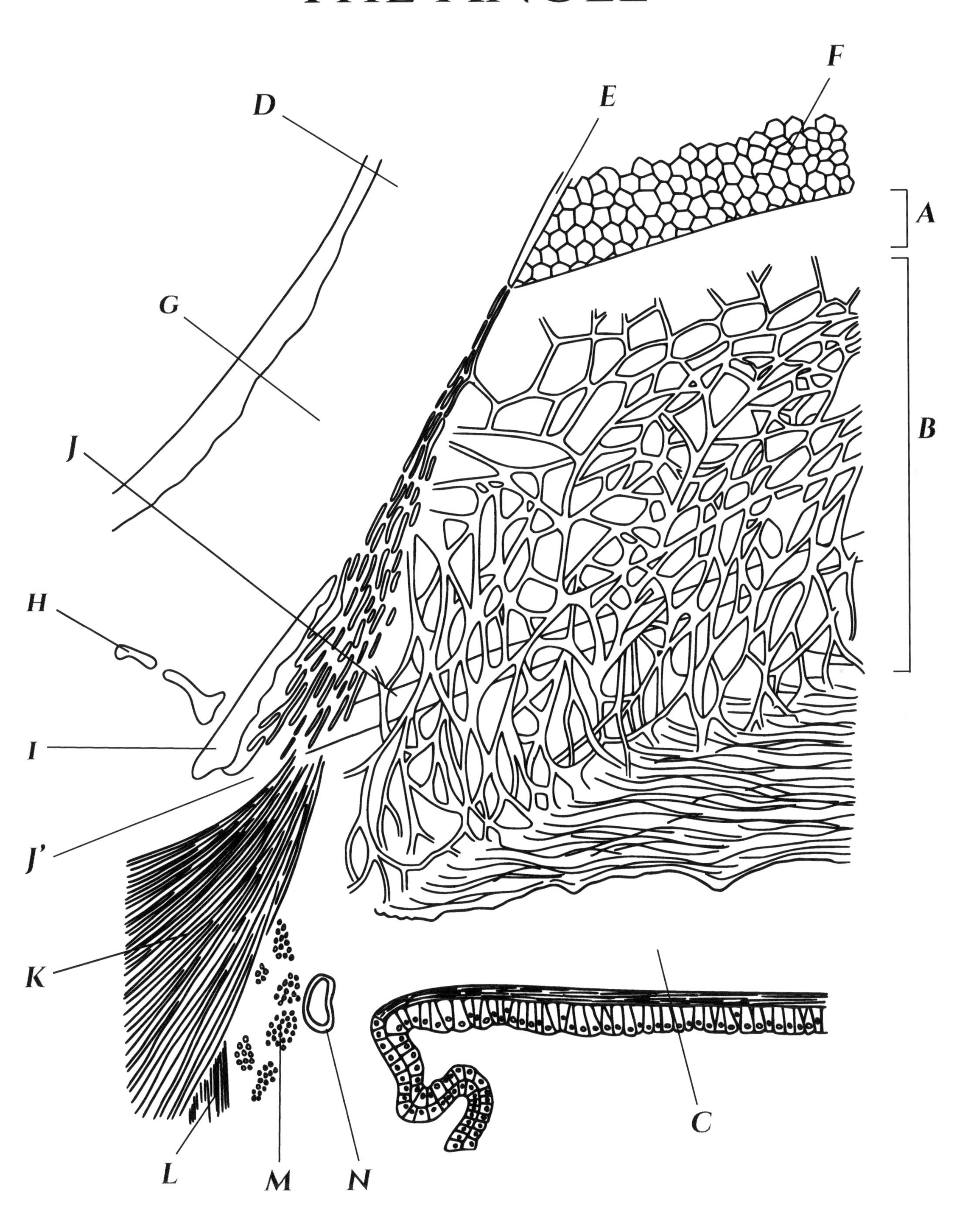

Aqueous Drainage

Aqueous humor is continuously secreted by the ciliary epithelium into the posterior chamber. The aqueous circulates through the pupil and into the anterior chamber. From the anterior chamber, the aqueous drains through two routes, the convention route through the trabecular meshwork and Schlemm's canal, and the unconventional uveoscleral route. Approximately 70-95% of aqueous humor leaves the anterior chamber via the conventional trabecular route, and 5-30% leaves via the unconventional uveoscleral.

In the conventional route, drainage begins with filtration through the trabecular meshwork. After passing through the trabecular meshwork, aqueous humor is carried across the inner wall Schlemm's canal into the lumen. Schlemm's canal is a circumferential channel lined with endothelial cells. The endothelial cells of the inner wall faces the juxtacanalicular trabecular meshwork, where giant vacuoles transport the aqueous humor into Schlemm's canal.

From Schlemm's canal, aqueous humor can take one of two pathways. Some aqueous passes from Schlemm's canal to external collector channels that join episcleral veins, and some aqueous passes from Schlemm's canal to efferent vessels, which join deep scleral veins before joining episcleral veins. From the episcleral veins, aqueous will travel with venous blood to the anterior ciliary veins, to the muscular veins, and to the superior and inferior ophthalmic veins.

In the unconventional route (not shown), aqueous humor passes through the uveal meshwork at the anterior portion of the ciliary muscle. The aqueous passes into the suprachoroidal space and joins the venous drainage of the eye.

Figure Description

Aqueous drainage from the anterior segment, the conventional outflow pathway, adapted from Hogan, Alvarado, Weddell, 1971

Key

A Schlemm's canal
B external collector channels
C deep scleral plexus
D intrascleral plexus
E aqueous vein

Aqueous Drainage

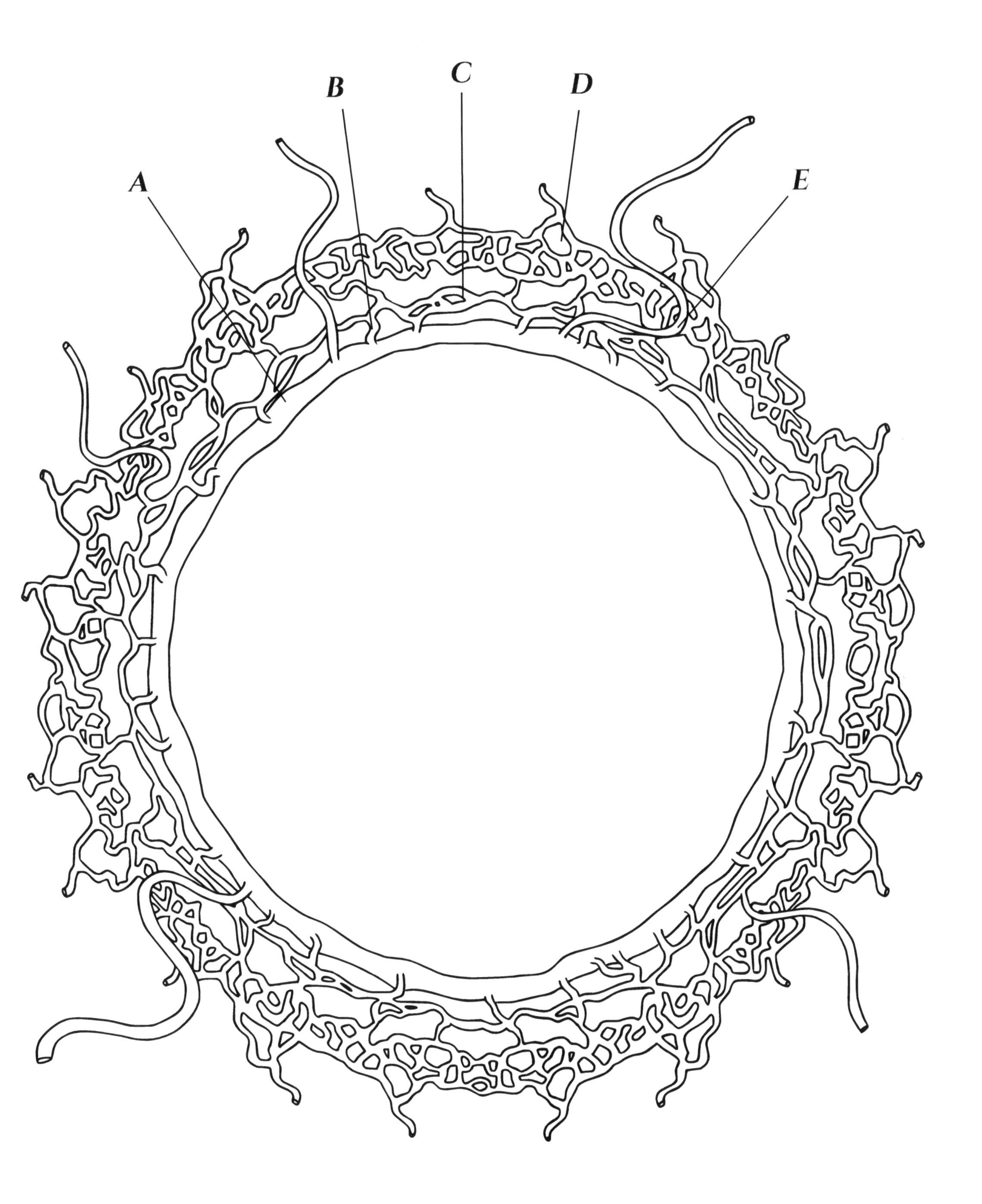

The Iris

The iris is a flat annular structure that is part of the uveal tunic, dividing the anterior segment into the anterior and posterior chambers. The iris is the "colored" part of the eye, with about 79% of the population having a brown iris color, followed by 8% blue, 10% hazel/amber, and 2% green. Every individual's iris is unique, even more unique than a fingerprint.

The iris functions to regulate the amount of light entering the eye and contribute to retinal image quality through control of pupil size. The iris also serves as an attachment for the trabecular meshwork and facilitates aqueous outflow by pumping aqueous into the angle with contraction and relaxation.

The average iris diameter is 12 mm, with a circumference of about 38 mm. The free edge of the iris surrounds the pupil, which is the aperture of the iris. The pupil is slightly nasal and inferior relative to the center of the iris. Pupil diameter ranges from 1.5 to 8 mm, depending on ambient illumination and autonomic tone. The attached edge of the iris to the ciliary body is the iris root. The iris is thinnest at the root, approximately 500 microns.

The collarette is a thickened circular ridge, approximately 1.5 mm from the pupil. The region of the iris spanning from the pupil to the collarette is the pupillary zone, and from the collarette to the iris root is the ciliary zone. The collarette is the thickest portion of the iris, approximately 600 microns in cross section.

The iris is highly vascularized. Two circumferential arterial circles supply the iris: the major arterial circle resides in the ciliary muscle, and the minor arterial circle resides in the collarette. Radially oriented vessels anastomose between the major and minor arterial circles and travel from the minor arterial circle to the pupil margin. These radial vessels are surrounded by connective tissue and appear as radially oriented ridges on the anterior surface of the iris. In the pupillary zone, these ridges are called trabeculae, and in the ciliary zone, they are called long radial ridges. Gaps in the anterior surface of the iris tissue, most commonly found near the collarette, are Fuch's crypts. These crypts allow aqueous humor to flow in and out of the iris stroma.

Figure Description

Isolated iris, anterior view

Key

A	pupil
B	pupillary ruff
C	trabeculae
D	collarette
E	contraction furrow
F	Fuch's crypt
G	long radial ridge

The iris contains two smooth muscles that control pupil size. The sphincter muscle is a circumferential muscle in the pupillary zone that acts to constrict the pupil through parasympathetic innervation. The dilator muscle is a radially oriented muscle spanning from the pupil to the iris root that acts to dilate the pupil through sympathetic innervation. As the pupil dilates, the iris tissue folds on itself, giving rise to contraction furrows that are seen in the ciliary zone.

THE IRIS

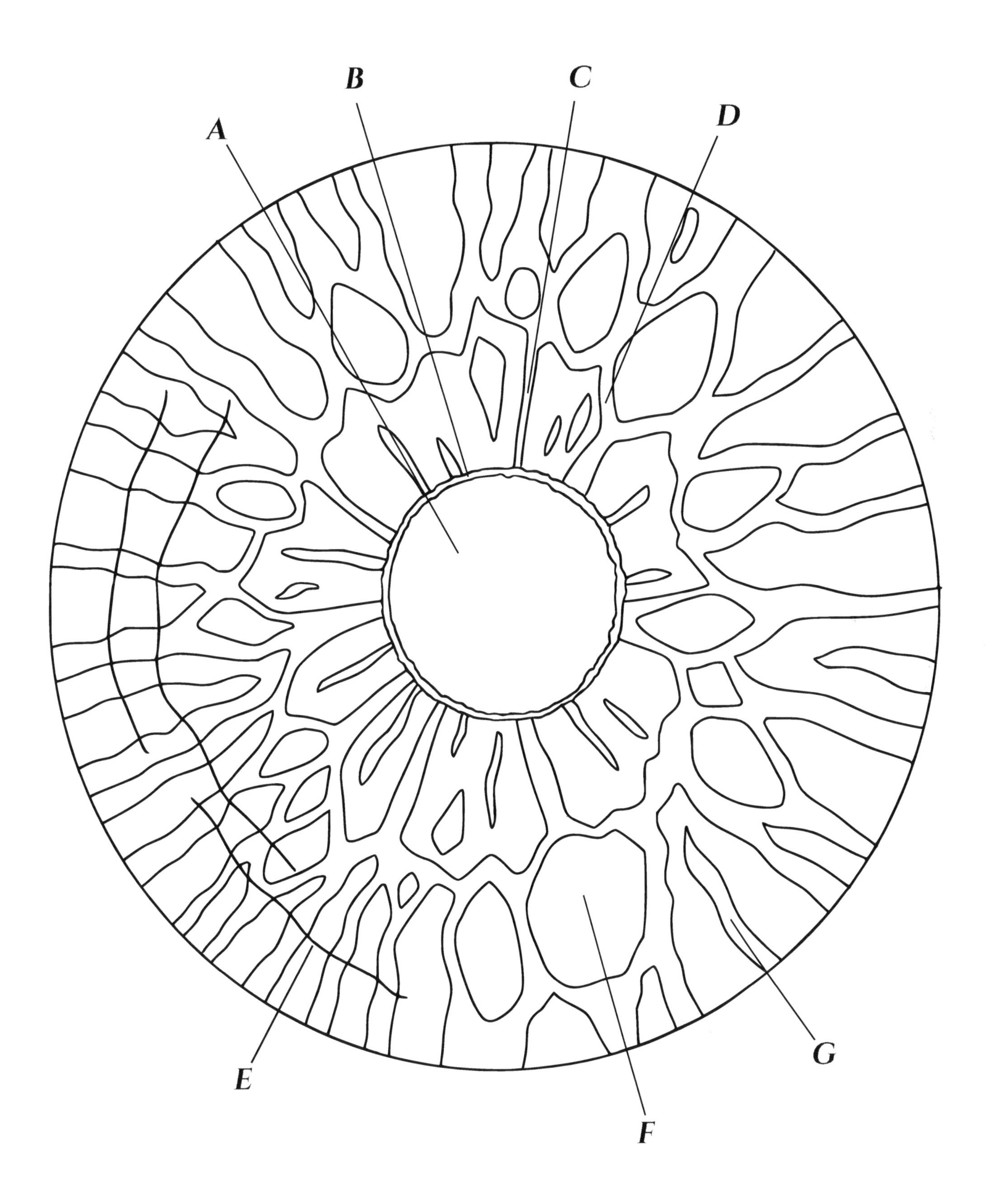

Iris Histology

The iris consists of four layers, including the anterior border layer, stroma, anterior epithelium, and posterior epithelium.

The anterior border layer is the most anterior layer of the iris. This layer is formed by a collection of fibroblasts (more superficial), melanocytes (deep to the fibroblasts), and collagen fibrils. The concentration of melanin within the melanocytes of this layer plays an important role in the color of the iris.

The stroma makes up most of the thickness of the iris. The stroma consists of loose collagenous tissue with numerous blood vessels and nerves. The stroma also contains fibroblasts and melanocytes, but less dense than in the anterior border layer, Lymphocytes, macrophages, mast cells, and clump cells are also found within the stroma. The sphincter muscle is located in the pupillary zone in the stroma.

Iris vasculature originates from the long posterior ciliary arteries, with some contribution from the anterior ciliary arteries. Vessels form the major arterial circle (located within the ciliary muscle), which gives off radial branches into the iris. The minor arterial circle runs within the collarette, and gives off branches that anastomose with radial branches from the major arterial circle. Branches also travel towards the pupil from the minor arterial circle. The walls of iris capillaries are non-fenestrated and have tight junctions, forming part of the blood-aqueous barrier.

The iris has sensory, sympathetic, and parasympathetic innervation. The sensory and sympathetic innervation are carried in both the long and short ciliary nerves, and the parasympathetic innervation is carried in only the short ciliary nerves.

The sphincter muscle is an 0.75-1 mm wide annular smooth muscle encircling the pupil, responsible for miosis, or pupil constriction. It is innervated by parasympathetic nerve fibers, carried from the oculomotor nerve (CN III). Preganglionic fibers travel from the Edinger-Westphal nucleus of the midbrain and synapse in the ciliary ganglion, and post-ganglionic fibers are carried in the short ciliary nerves to the sphincter muscle.

Posterior to the stroma are two layers of epithelium, an anterior myoepithelial layer and a posterior pigmented epithelial layer, described in the next figure.

Figure Description

Cross section of iris, pupillary zone

Key

A pupillary ruff
B anterior border layer
C stroma
D collarette region
E vasculature
F sphincter muscle
G anterior myoepithelium
H posterior pigmented epithelium

Iris Histology

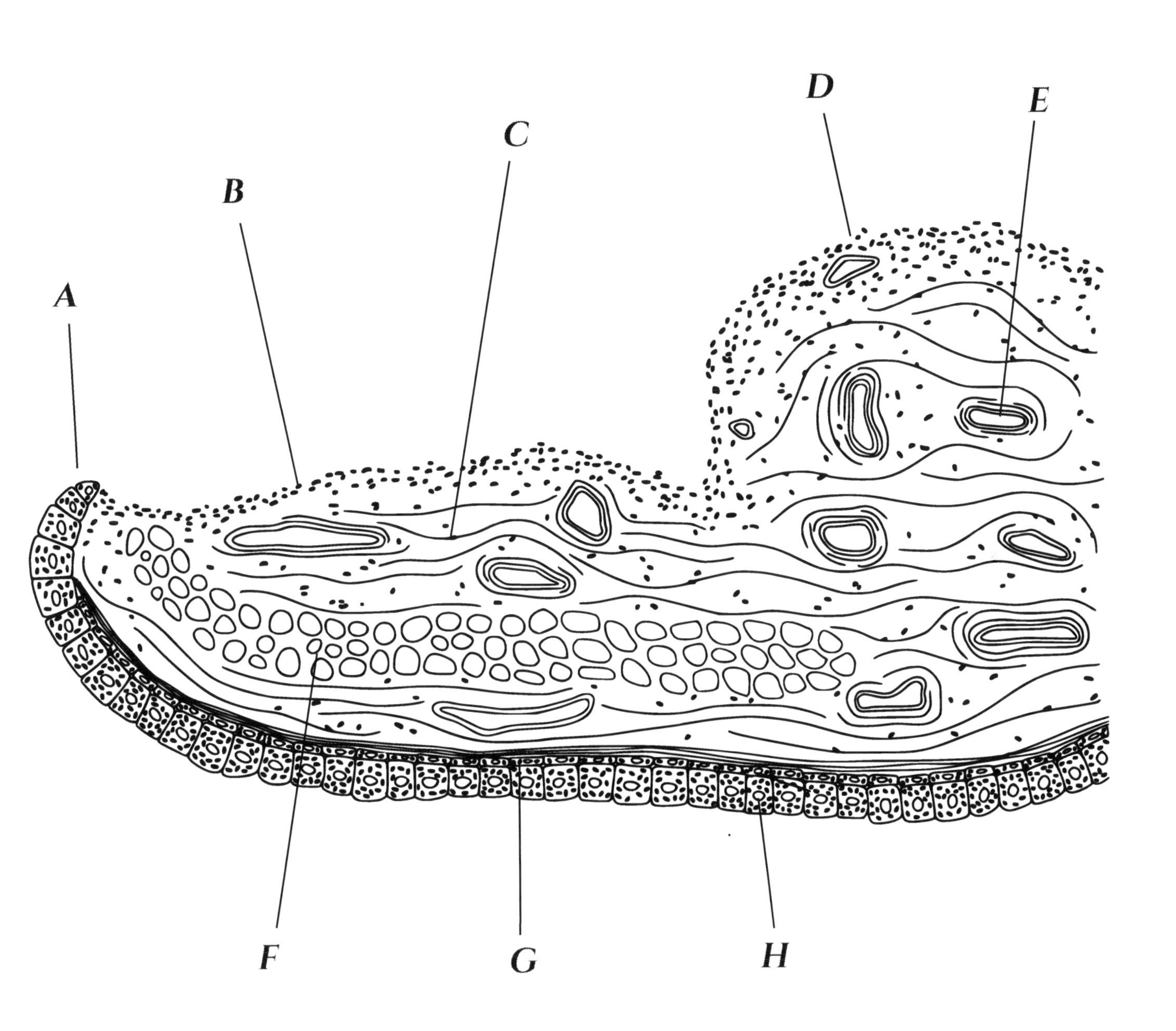

Iris Epithelium

The posterior surface of the iris is lined by two layers of epithelium, the anterior myoepithelium and posterior pigmented epithelium. These two layers are situated apex to apex, and joined together with desmosomes. Cells are joined to neighboring cells through desmosomes and tight junctions on their lateral walls.

The anterior epithelial layer spans from the pupillary zone to the ciliary body, where the cells transition into the pigmented epithelium of the ciliary body. The anterior epithelium is a myoepithelial layer; the more posterior apical side of the cells is epithelial-like, whereas the anterior basal portion is muscle-like, constituting the dilator muscle. The apical portion consists of a single layer of pigmented cuboidal cells. The basal processes project into the stroma and are joined by tight junctions; they are radially oriented and directed towards the pupil.

The dilator muscle is a smooth muscle that is responsible for mydriasis, or pupil dilation. It is innervated by sympathetic nerve fibers. Information is carried from the midbrain to the lateral horn of the thoracic spinal cord. Pre-ganglionic fibers exit the spinal cord and enter the sympathetic chain, synapsing in the superior cervical ganglion. Post-ganglionic fibers travel to the eye via the carotid plexus, then fibers join the ophthalmic nerve (CN V1) and enter the eye with the long and short ciliary nerves. Some sympathetic fibers travel through the ciliary ganglion, but do not synapse there.

The posterior epithelium is a densely pigmented layer of columnar cells that extends from the pupillary margin and into the ciliary body, where it transitions into the non-pigmented epithelium of the ciliary body. At the pupil margin, the posterior epithelium extends around the pupil edge to meet the anterior border layer, forming a pigmented ring at the pupil margin known as the pupillary ruff. The basement membrane of the posterior epithelium faces the posterior chamber; it is continuous with the basement membrane of the non-pigmented epithelium of the ciliary body.

Figure Description

Epithelial layers of the posterior iris

Key

A basal muscle-like anterior epithelium
B apical epithelial-like anterior epithelium
C posterior epithelium
D basement membrane
E desmosome
F pigment granule (melanin)
G tight junction

Iris Epithelium

and dilator muscle

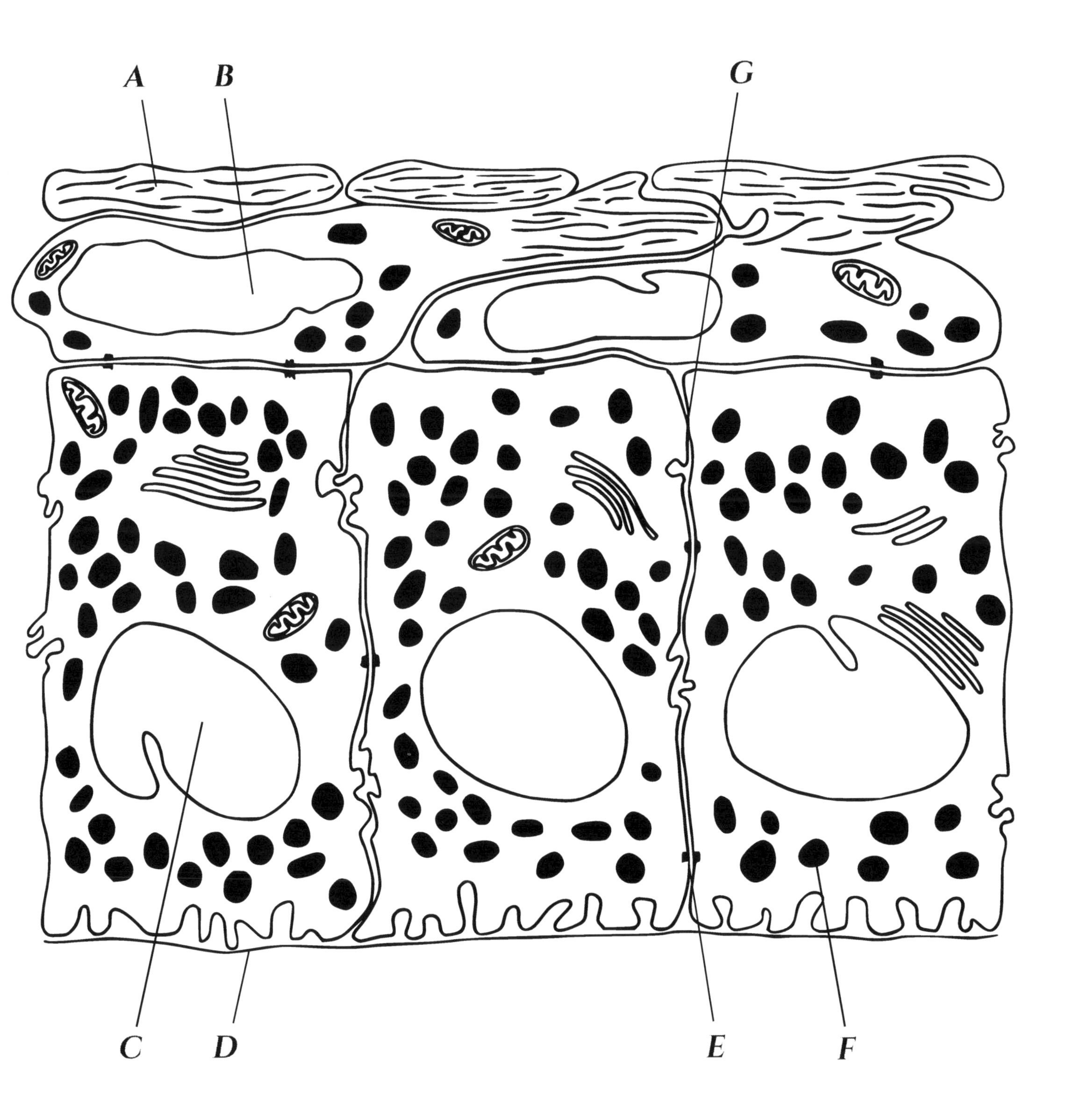

The Ciliary Body

The ciliary body is a circumferential structure located peripheral to the iris root. The ciliary body consists of the ciliary muscle, which is responsible for accommodation (focusing of the eye), and the ciliary epithelium, which produces and secretes aqueous humor. It is part of the uveal tract, being richly vascularized and highly pigmented.

The ciliary body faces the sclera externally and faces the posterior chamber and vitreous internally. The external-most layer is the supraciliaris, consisting of loose connective tissue that is continuous with the suprachoroidia of the choroid. Nerves and vessels travel within the supraciliaris. Moving more internally, the next layer is the ciliary muscle, then a stromal layer with blood vessels surrounded by a basal lamina, then two epithelial layers and an internal limiting membrane.

The ciliary body is divided in two regions; the pars plicata region, aka corona ciliaris, is more anterior and consists of the ciliary processes, and the pars plana region, aka orbicularis ciliaris, is more posterior and flat, extending to the ora serrata. The ora serrata is the transition of ciliary body to retina. The ora serrata has a characteristic scalloped appearance, forming dentate processes.

The blood supply to the ciliary body primarily originates from the long posterior ciliary arteries. Venous drainage is via the vortex veins

The ciliary ring is formed by the annulus of ciliary processes. The crystalline lens is suspended within the ciliary ring by zonular fibers, which span from the ciliary processes to the lens equator. The region between the tips of the ciliary processes and the lens equator is the circumlental space, which is also known as the canal of Hanover. Just anterior to the zonular fibers is the posterior chamber proper, and just posterior to the zonular fibers is the canal of Petit, which is the retrolental space between the posterior zonular fibers and the anterior face of the vitreous.

Figure Description

Posterior view of eye, bisected at equator and posterior segment removed, adapted from Gray, 1918

Key

A ora serrata
B pars plana of ciliary body
C pars plicata of ciliary body
D zonular fibers spanning the ciliary space
E crystalline lens equator
F posterior surface of lens

The Ciliary Body

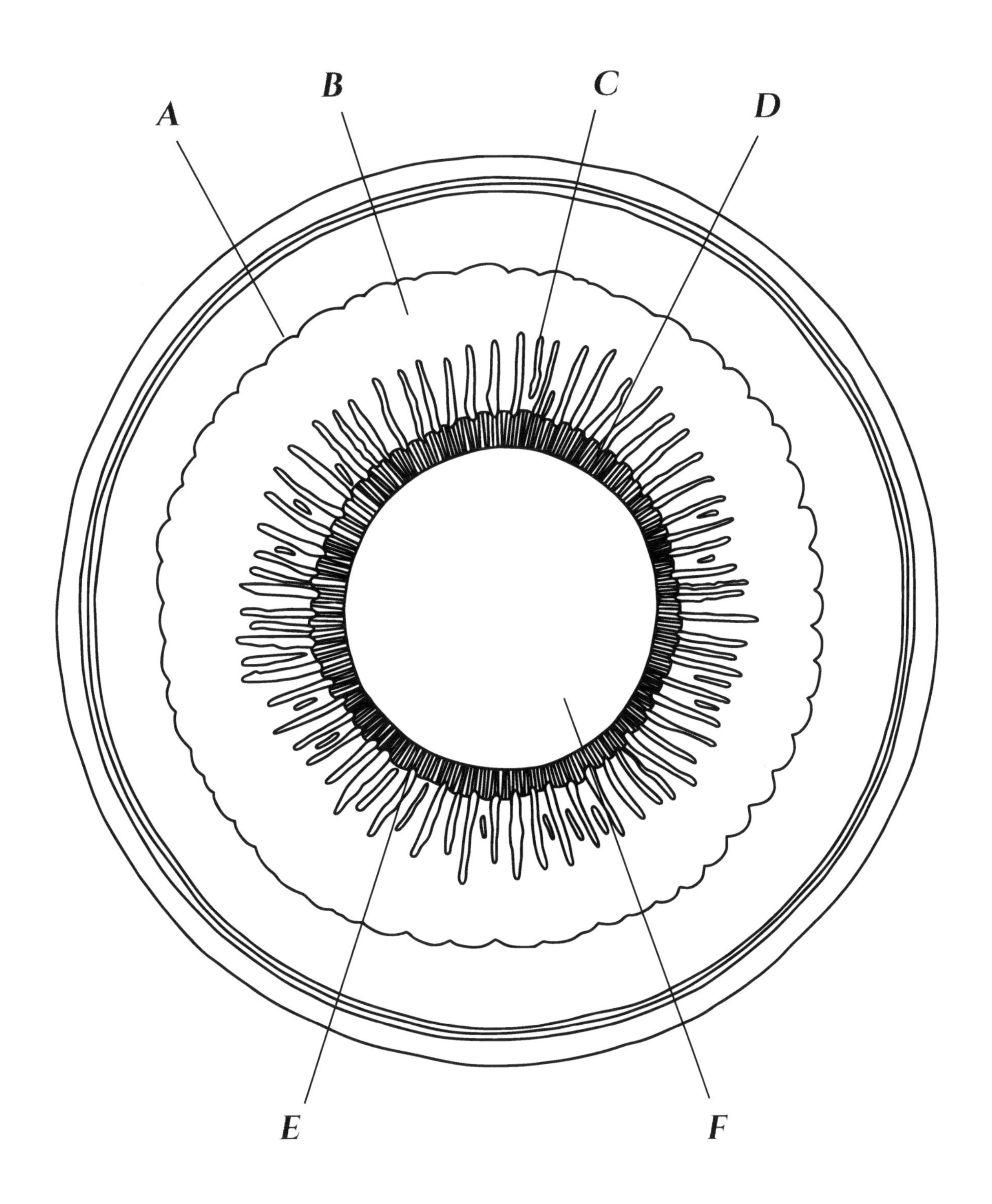

The Ciliary Muscle

The ciliary muscle within the ciliary body is a multi-unit smooth muscle consisting of longitudinal, radial, and circular muscle fibers. The ciliary muscle is parasympathetically innervated. Preganglionic fibers travel from the Edinger-Westphal nucleus of the midbrain with the oculomotor nerve (CN III) and synapse in the ciliary ganglion, and post-ganglionic fibers are carried in the short ciliary nerves to the ciliary muscle. With contraction, the ciliary muscle moves anteriorly and centripetally, releasing resting tension on the zonular fibers and allowing the crystalline lens to take a more spherical, accommodated form.

The longitudinal muscle fibers, aka Brucke's muscle, are most externally located and make up the largest portion of the ciliary muscle. The longitudinal fibers originate at the scleral spur and trabecular meshwork and extend posteriorly towards the choroid. They are oriented parallel to the sclera.

The radial muscle fibers are the middle layer of the ciliary muscle, and are oriented in a criss-cross arrangement.

The circular muscle fibers, aka Müller's muscle, are most internal and arranged circumferentially. The major arterial circle of the iris is generally found within the circular muscle fibers.

Figure Description

Anterior cornea and sclera removed to reveal angle and ciliary muscle, adapted from Hogan, Alvarado, Weddell, 1971

Key

A	posterior cornea
B	trabecular meshwork, anterior view
B'	trabecular meshwork, cross section
C	external collector channel
D	Schlemm's canal
E	circular fibers of ciliary muscle
F	radial fibers of ciliary muscle
G	longitudinal fibers of ciliary muscle
H	pars plana of ciliary body
I	ciliary process
J	iris
K	scleral spur

The Ciliary Muscle

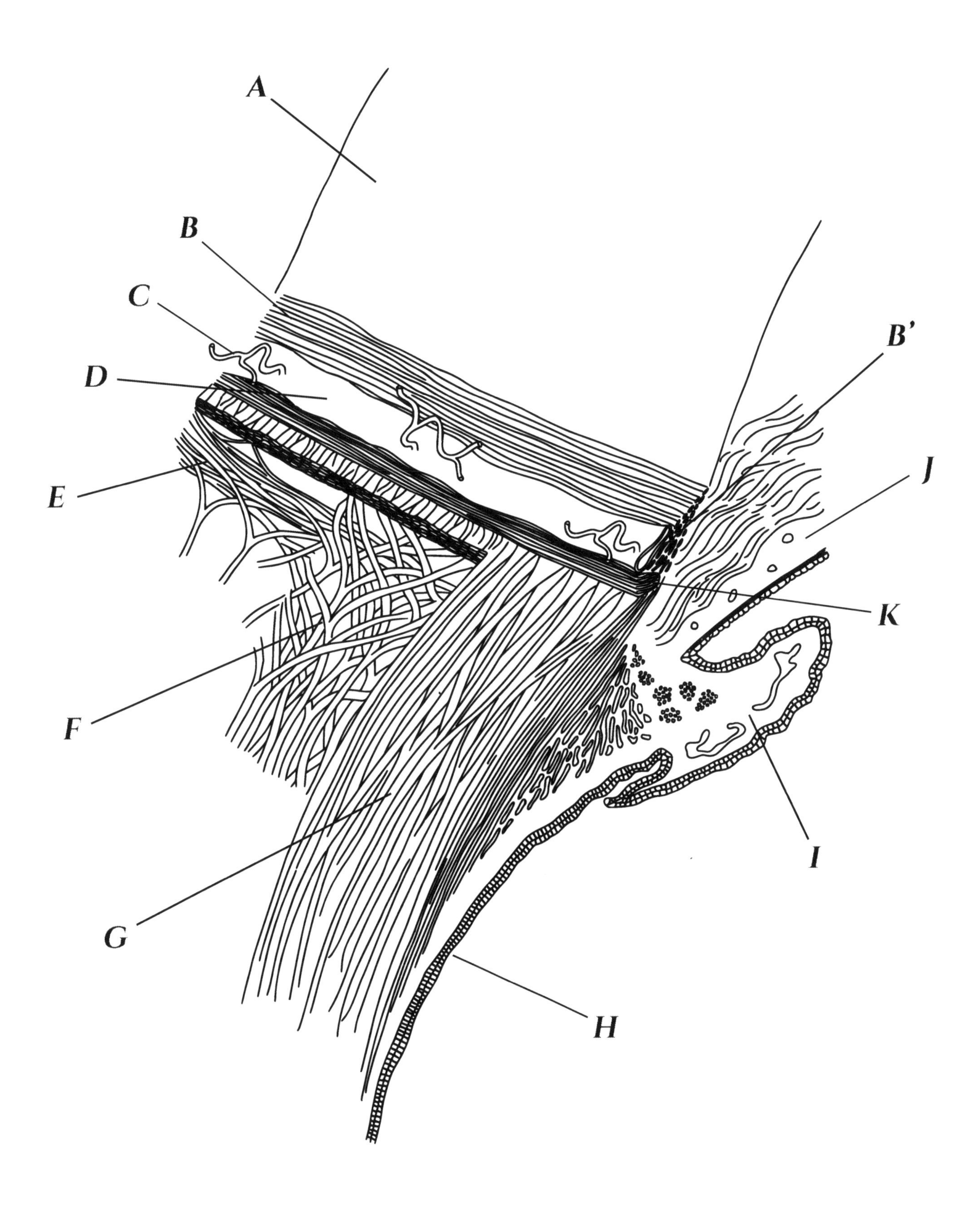

A Ciliary Process

The pars plicata region of the ciliary body consists of 70-80 ciliary processes which are finger-like extensions into the posterior chamber. The ciliary processes are the main site of aqueous humor production and secretion. The ciliary processes also serve to anchor the zonular fibers that suspend the crystalline lens.

Each ciliary process consists of a core of blood vessels surrounded by connective tissue and a double layer of epithelium. The blood vessels are fenestrated and leaky. Plasma passes out of the vessels into the stroma, where the bilayered epithelium forms a functional blood-aqueous barrier, secreting aqueous humor into the posterior chamber.

The bilayered epithelium consists of a pigmented epithelial layer (facing the stroma) and non-pigmented epithelial layer (facing the posterior chamber). The epithelial layers are organized apex to apex. Aqueous humor forms as fluid moves from the capillary network across (or between) the epithelial cells and into the posterior chamber. Fluid moves via 1) diffusion, 2) ultrafiltration, and 3) active secretion.

The pigmented epithelium consists of a single layer of cuboidal cells, with the basement membrane attached to the basal lamina of the ciliary body stroma. Anteriorly, the pigmented epithelium of the ciliary body is continuous with the anterior myoepithelium of the iris. Posteriorly, it is continuous with the retinal pigment epithelium.

The non-pigmented epithelium is a single layer of cuboidal/columnar cells. Anteriorly, the non-pigmented epithelium of the ciliary body is continuous with the pigmented posterior epithelium of the iris. Posteriorly, it transitions into the neural retina at the ora serrata.

Figure Description

Histology of a ciliary process

Key

A arteriole
B stroma
C pigmented epithelium
D nonpigmented epithelium
E internal limiting membrane

A Ciliary Process

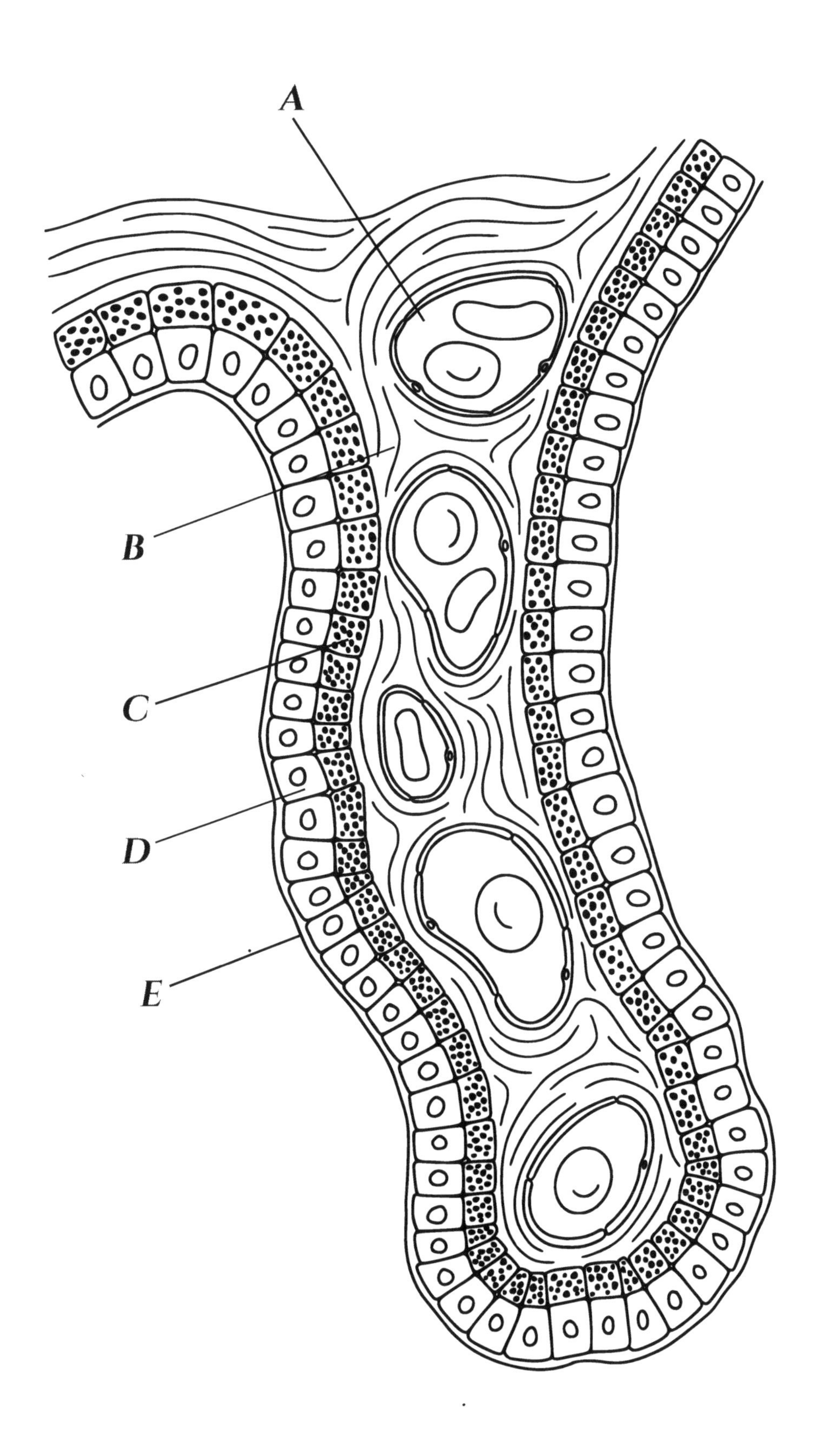

Zonular Fibers

Zonular fibers, aka zonules of Zinn, are thin fibers that suspend the crystalline lens within the ciliary ring. The zonular fibers primarily originate from the basement membrane of the non-pigmented epithelium in the pars plana region of the ciliary body. There are four groups of zonular fibers; 1) anterior, equatorial, and posterior fibers, 2) tension fibers, 3) hyaloid fibers, and 4) narrow bands of fibers.

The anterior, equatorial, and posterior zonular fibers extend from the ciliary processes to the lens equator. The anterior fibers insert into the lens capsule anterior to the equator, and the posterior fibers insert posterior to the equator. The spaces between these fibers are filled with aqueous humor, and form the canal of Hannover. Posterior to the posterior zonular fibers is the canal of Petit.

The tension zonular fibers extend from the ciliary processes posteriorly, over the pars plana region of the ciliary body.

The hyaloid zonular fibers span from the pars plana to the posterior surface of the lens, following along the anterior face of the vitreous.

The narrow bands of zonular fibers span across the posterior surface of the lens.

Figure Description

Posterior view of ciliary body and crystalline lens, adapted from Hogan, Alvarado, Weddell, 1971

Key

A iris
B crystalline lens
C crystalline lens equator
D anterior zonular fibers
E equatorial zonular fibers
F posterior zonular fibers
G ciliary process
H tension zonular fibers
I ora serrata
J circumlental space, aka canal of Hannover
K pars plicata of ciliary body
L pars plana of ciliary body

Zonular Fibers

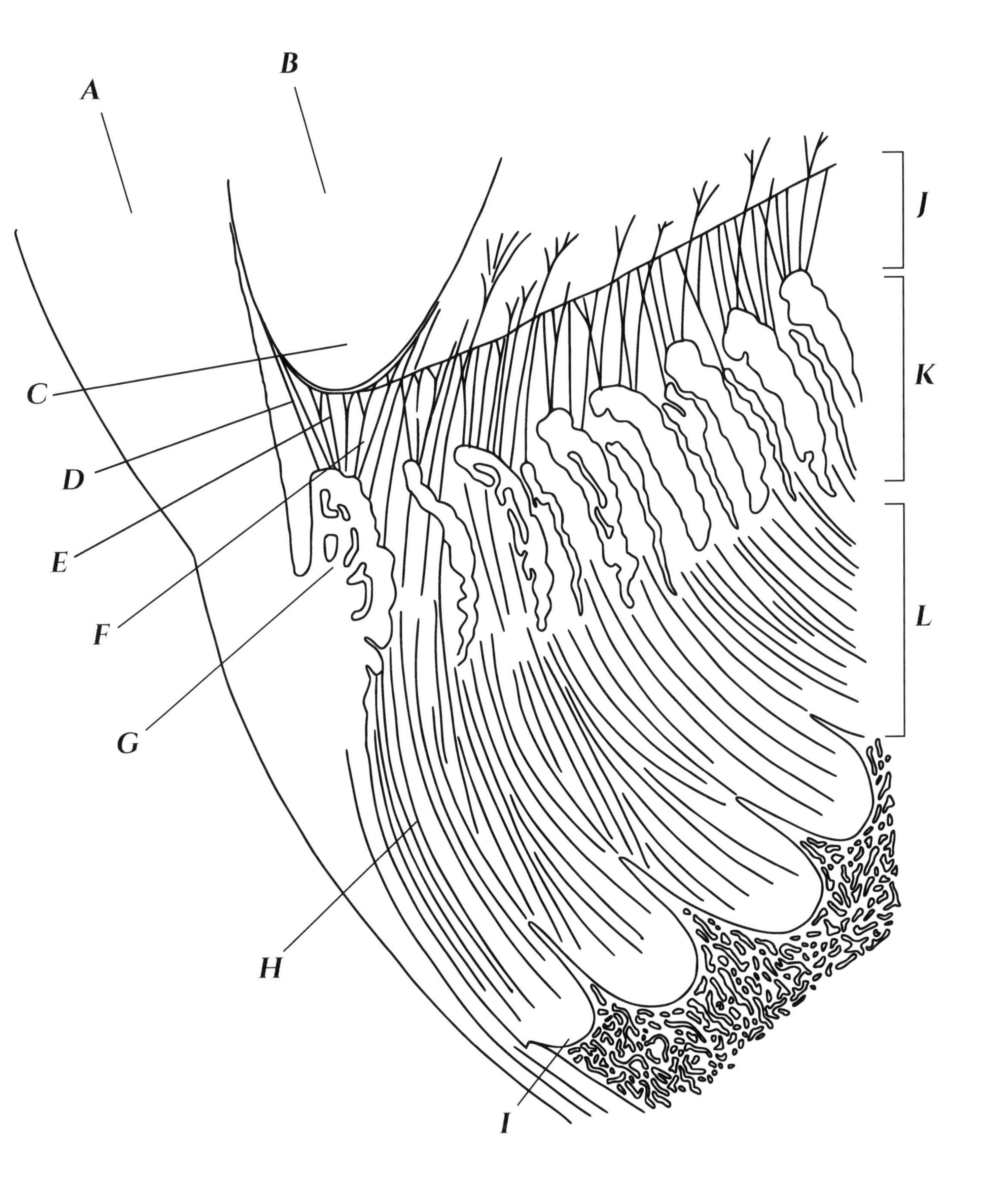

The Crystalline Lens

The crystalline lens is a transparent, avascular biconvex structure in the posterior chamber, situated between the iris and vitreous. The functions of the lens are to transmit visible light, absorb ultraviolet light, and refract light. The lens contributes 18-20 D of refracting power to the eye, which can increase to over 30 D in a young accommodating eye focusing on a near object.

The lens continues to grow throughout life. In an adult eye, the diameter of the lens is about 10 mm. The center point of the anterior lens surface is the anterior pole, and of the posterior surface is the posterior pole. Connecting the poles, the axial thickness of the lens is approximately 3.5-4 mm. The anterior surface curvature is flatter than the posterior surface. Both surfaces are aspheric, being steeper in the center and flattening towards the periphery. The anterior radius of curvature is 10 mm, and the posterior radius of curvature is 6 mm. In an anterior or posterior view, the outer edge of the lens is the equator.

The index of refraction of the lens ranges from 1.38 in the periphery to 1.41 in the center. This distribution of refractive index is referred to as the gradient refractive index, or GRIN. The lens transmits visible light and acts as an ultraviolet filter. The molecule 3-hydroxykynurenine absorbs ultraviolet light. Over time, this leads to oxidative damage, yellowing, and opacification. The lens is most transparent and malleable in childhood, then begins to opacify and harden with age, contributing to the formation of cataracts and loss of accommodative ability, aka presbyopia.

The lens is completely surrounded by the lens capsule, an acellular basement membrane of collagen type IV and glycosaminoglycans (GAGs). The lens capsule forms a barrier, preventing large molecules, bacteria, and inflammatory cells from entering the lens. The lens capsule also helps to mold the lens substance into a more spherical shape during accommodation. This shape change is accomplished in part by thickness variations in the capsule. The capsule is thinnest at the poles and equator, and thickest in the mid-periphery, which helps to increase axial thickness during accommodation.

Deep to the lens capsule on the anterior surface of the lens is a single layer of epithelial cells. Towards the anterior pole, the epithelial cells are cuboidal, and near the equator, the cells become columnar.

Figure Description

Sagittal cross section of crystalline lens

Key

- A anterior pole
- B posterior pole
- C equator
- D central zone of anterior epithelium
- E germinative zone of anterior epithelium
- F equatorial zone of anterior epithelium
- G anterior capsule
- H lens bow
- I posterior capsule
- J cortex
- K adult nucleus
- L fetal nucleus
- M embryonic nucleus

The epithelium is divided into three regions: 1) the central zone at the anterior pole, 2) the germinative zone in the mid-periphery, where cells undergo mitosis, and 3) the equatorial zone, where cells transition into lens fibers.

Lens epithelial cells are never shed because the lens capsule surrounding the lens keeps the substance contained. Rather, epithelial cells transition into new lens fibers, which are continuously laid down during life, compressing earlier generations of lens fibers towards the center of the lens. Epithelial cells migrate to the equatorial zone, where they elongate, rotate their axis, and lose their nucleus. The region of newly forming lens fibers that still have a nucleus form the nuclear bow. Once the nucleus is lost, the cell is a true lens fiber.

The majority of the lens substance is made up of lens fibers, with the oldest lens fibers in the center of the lens. These lens fibers make up the embryonic nucleus. Surrounding the embryonic nucleus is the fetal nucleus, which is surrounded by the adult nucleus. The outermost lens fibers surrounding the adult nucleus make up the lens cortex.

As new lens fibers extend from the equator towards the poles, they meet fibers growing from the opposite direction. The fiber meet and form a suture, which takes on a characteristic Y shape in the anterior lens, and an inverted Y in the posterior lens.

The Crystalline Lens

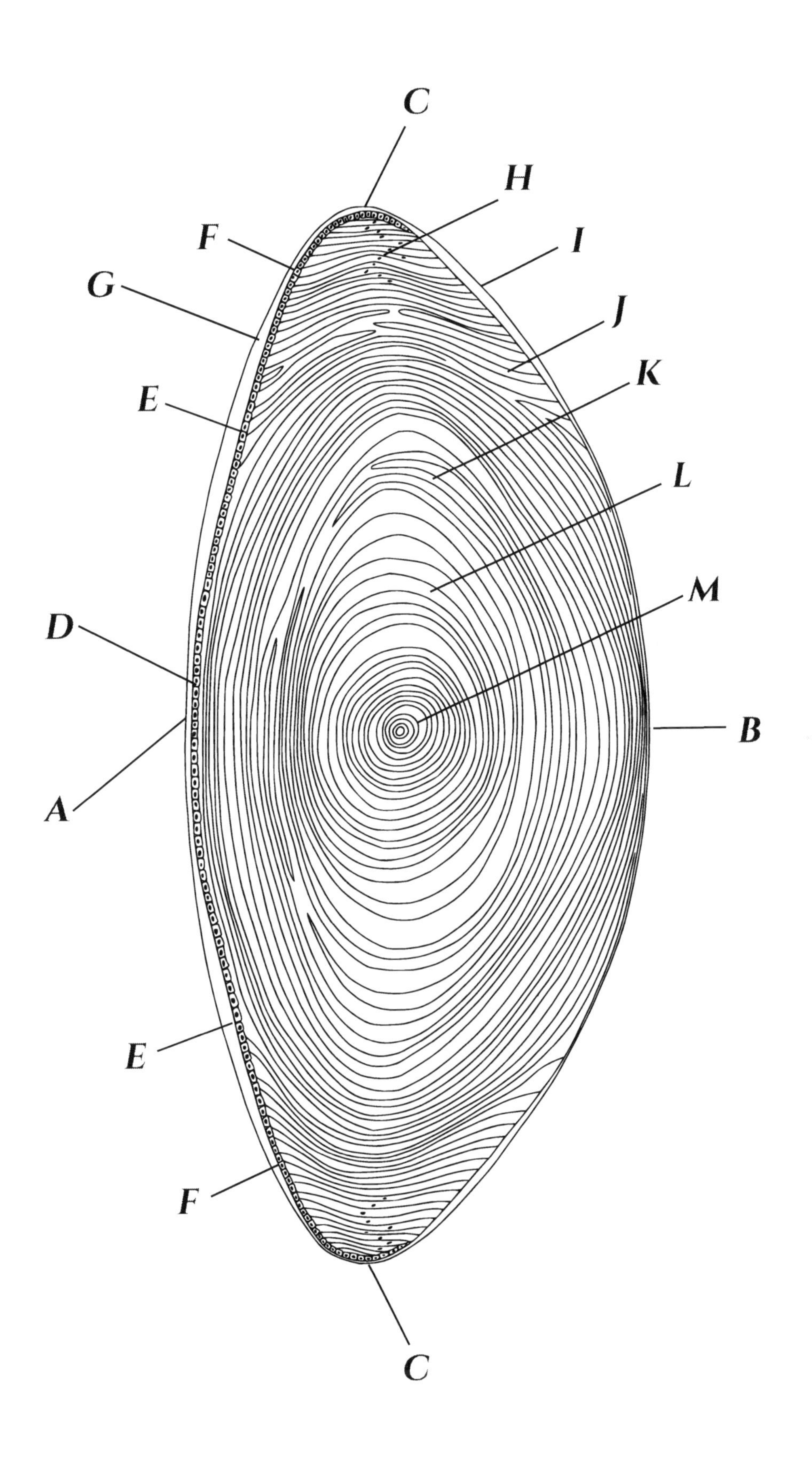

Crystalline Lens Histology

The outermost layer of the crystalline lens is the acellular lens capsule, which envelopes the entire lens. The next layer is a single layer of epithelial cells, found only on the anterior surface. There is no epithelium on the posterior lens. The remaining lens substance consists of lens fibers.

Lens epithelial cells form a single layer over the anterior surface of the lens, just deep to the capsule. The apical surface faces the capsule and the basal surface faces the lens fibers. These epithelial cells contain gap junctions, sodium potassium pumps, ion channels, and aquaporins.

The lens is composed of 60-70% water and 30-40% proteins. About 90% of the proteins are crystallins, specifically, alpha, beta, and gamma crystallins. Alpha crystallin is considered to be a molecular chaperone because it helps the other proteins recover from injury.

The lens also contains a high concentration of glutathione, which helps to protect the lens from oxidative damage.

Lens fibers are joined together by interlocking ball-and-socket and tongue-and-groove joints. This arrangement allows the lens to change shape during accommodation, but maintain connections with neighboring lens fibers. Gap junctions, ion channels, and aquaporins are also found in the lens fibers.

Figure Description

Histology of the anterior crystalline lens, adapted from Hogan, Alvarado, Weddell, 1971

Key

A	capsule
B	epithelial cell nucleus
C	gap junction
D	lens fibers
E	ball and socket joints

Crystalline Lens Histology

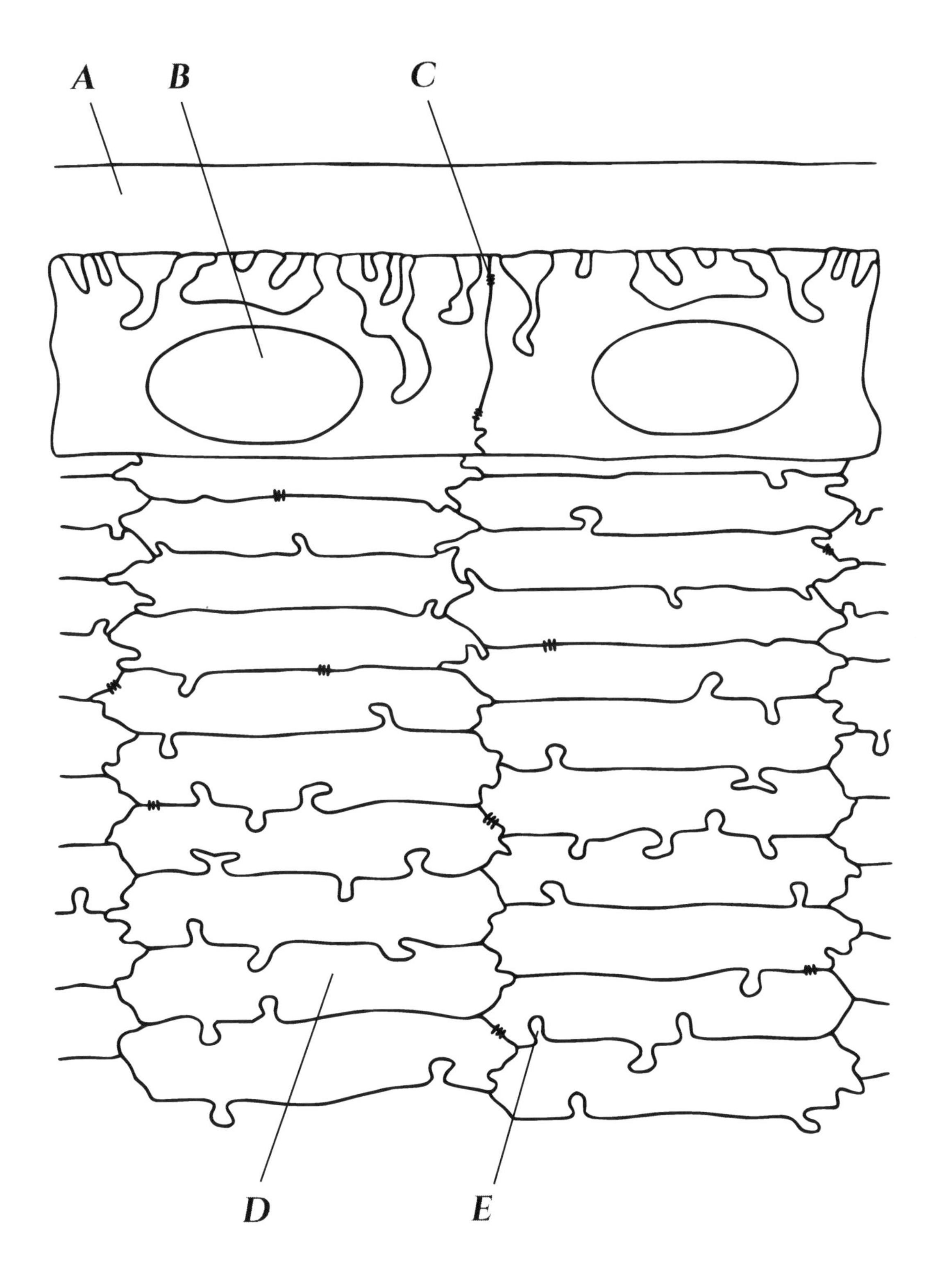

The Posterior Segment

Tenon's Capsule

Tenon's capsule, or fascia bulbi, is a thin transparent connective tissue layer just external to the episclera and sclera. Tenon's capsule spans from 1-2 mm posterior to the limbus, where it lays under the bulbar conjunctiva, to the optic nerve. Tenon's capsule separates the eye from the intraconal orbital fat.

Tenon's capsule is penetrated by the optic nerve, ciliary nerves and arteries, vortex veins, and tendons of the extraocular muscles. Tenon's capsule forms sheaths, or sleeves, surrounding the extraocular muscle insertion into the sclera. These connective tissue sleeves form muscle pulleys that prevent the rectus muscles from slipping along the globe during eye movements. The position of the rectus muscle's pulley is stable during ocular movements perpendicular to the axis of this muscle, and the position of the pulley moves with the eye during movements along the rectus muscle's axis.

The potential space between Tenon's capsule and the episclera is sub Tenon's space. Sub Tenon's space can be utilized for injection of anesthesia for ocular surgery or slow release medications.

Figure Description

Tenon's capsule of the right eye, isolated from the globe with extraocular muscles in place, adapted from Poirier and Charpy, 1912

Key

A Tenon's capsule
B superior rectus muscle
C superior oblique muscle
D lateral rectus muscle
E medial rectus muscle
F inferior rectus muscle
G inferior oblique muscle
H opening for optic nerve
I opening for superior temporal vortex vein
J opening for inferior temporal vortex vein

Tenon's Capsule

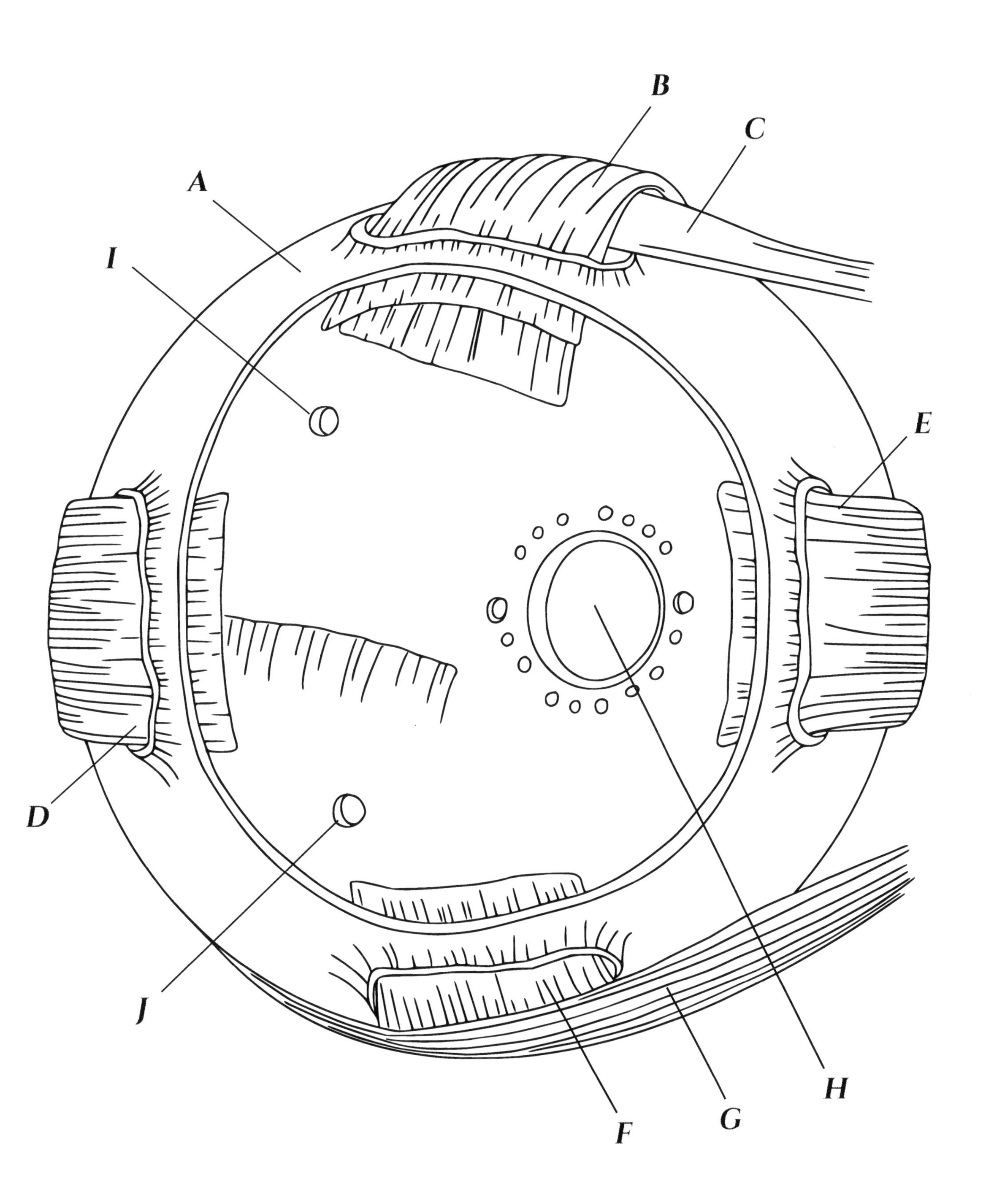

The Vitreous Humor

The vitreous humor is the largest structure of the eye, with a volume of about 4.4 mL, taking up 80% of the internal volume of the eye. The vitreous is a transparent gel-like tissue that occupies the vitreous chamber of the posterior segment. The vitreous functions to transmit light, store metabolites of the lens and retina, provide an avenue for movement of substances, and act as a shock absorber, protecting the retina during rapid eye movements and physical activity.

In an adult eye, the vitreous chamber is 15-16 mm in length. The outer region of the vitreous is the cortex, and the central region is the nucleus. The vitreous has a refractive index of 1.335. It transmits 90% of wavelengths from 300 to 1400 nm. The vitreous is most viscous at birth, being 100% gel, and liquefies with age, such that by 70 years of age, the vitreous is about 50% gel and 50% liquid.

The vitreous is composed of about 99% water with a collagen type II and hyaluronic acid matrix. Hyaluronic acid is produced by hylocytes in the vitreous cortex. Hyaluronic acid is a macromolecule of glycosaminoglycans (GAGs) that serves to maintain spacing between collagen fibrils. Fibroblasts and macrophages are also found in the vitreous.

The anterior surface of the vitreous is called the vitreous face, or anterior hyaloidia. There is a depression in the central region of the vitreous face, the patellar fossa, where the lens is found.

There are several regions where the vitreous is adhered to surrounding structures. Within the patellar fossa, there is a region of adherence of the vitreous to the mid-peripheral surface of the lens, known as the hyaloideocapsular ligament of Wieger. However, the vitreous is not adhered to the lens at the posterior pole, creating a potential space, known as Berger's space.

The vitreous base is a region that is adhered to the wall of the eye at the ora serrata. The vitreous base spans from about 2 mm anterior to 2 mm posterior to the ora serrata. The vitreous is also adhered to the retina around the optic disc, known as the area of Martigiani, and around the macula. As the vitreous liquefies with age, it can pull away from the retina. The posterior vitreous pulling away from the retina is a posterior vitreous detachment. While generally benign, it can result in the perception of flashes and floaters and, in some cases, a retinal detachment.

Figure Description

Isolated vitreous humor

Key

A	anterior vitreous face
B	patellar fossa
C	vitreous base
D	area of Martigiani

The Vitreous Humor

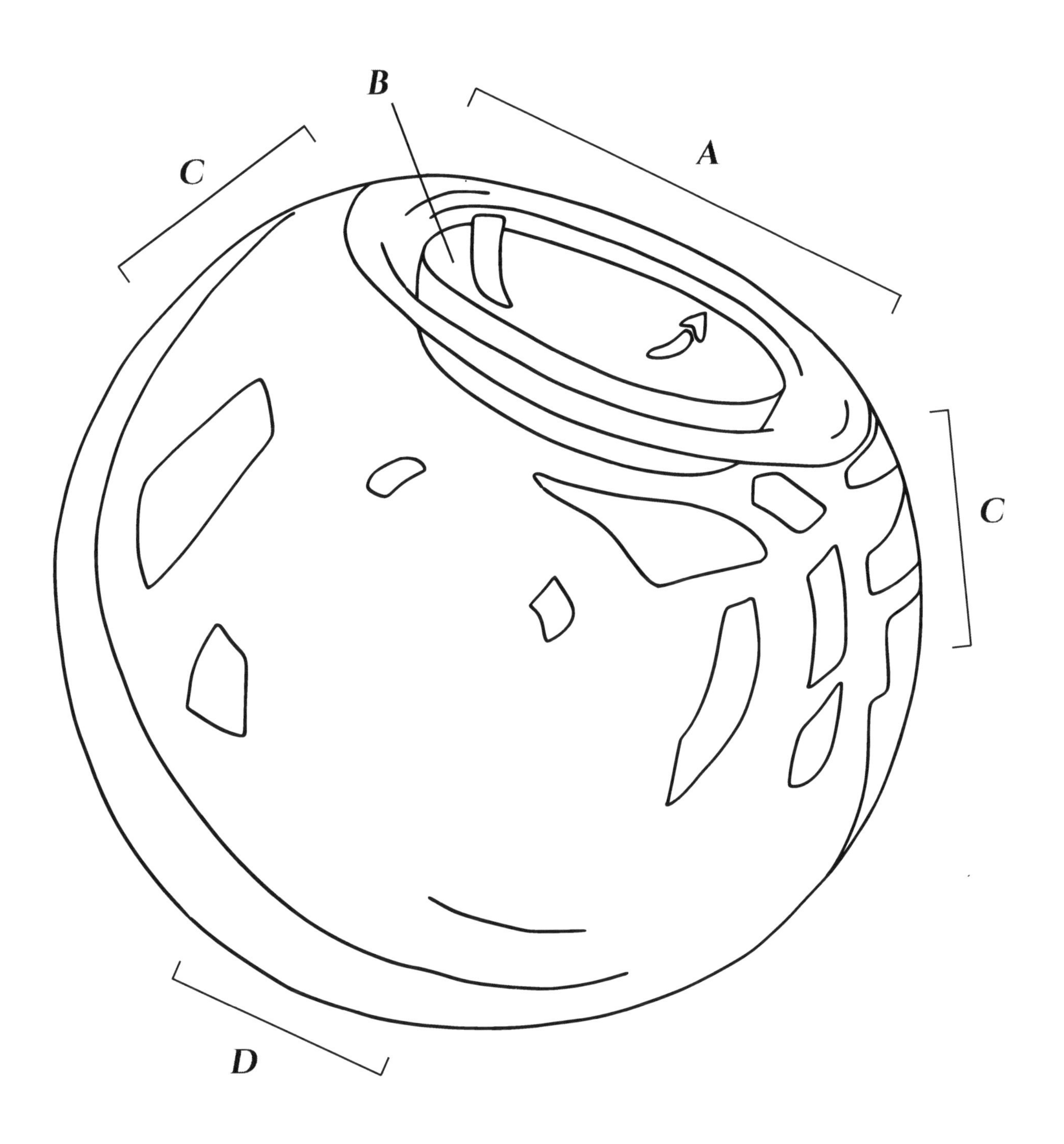

The Sclera and Choroid

The sclera is a white opaque collagenous tissue that forms the posterior 5/6 of the outer tunic of the eye. The sclera functions to maintain the shape of the eye. It is biomechanically strong, protecting the intraocular structures. The sclera extends from the limbus to the optic nerve. At the optic nerve head, the sclera is continuous with the lamina cribrosa. The mean radius of curvature of the sclera is 11.5 mm. Thickness ranges from 0.3 to 1.0 mm, being thickest at the posterior pole and thinnest at the insertion of the extraocular muscles. The sclera is primarily avascular, with a minimal blood supply from the episcleral vessels and choroid. Sensory innervation from the sclera is carried in the long and short ciliary nerves.

The sclera is primarily composed of dense, irregular collagen type I. The sclera consists of an outermost episclera, the middle sclera proper, and the innermost lamina fuscia. The episclera is loose connective tissue that is highly vascularized. The sclera proper is dense connective tissue that is primarily avascular. The lamina fuscia is a transition layer to the choroid, containing elastin fibers and melanocytes.

The choroid is the vascular layer between the sclera and retina, making up the posterior portion of the uveal tunic of the eye. The functions of the choroid are to provide nourishment to the sclera and outer retina, remove waste from the eye, regulate ocular temperature, absorb stray light, and serve as a pathway for vessels and nerves traveling through the eye.

The vasculature of the choroid supplies the sclera and the outer retina. The arterial vessels of the choroid originate from the long and short posterior ciliary arteries. The anterior ciliary arteries also provide a minor contribution to the anterior portion of the choroid. Venous drainage of the choroid is via the vortex veins.

The choroid contains connective tissue and is highly pigmented. The outermost layer of the choroid is the suprachoroidia, a transition layer from sclera to choroid, that is composed of loose connective tissue with melanin and fibroblasts. The next layer of the choroid is the vascular stroma, which consists of three layers with progressively smaller diameter blood vessels. Haller's layer has the largest diameter blood vessels, Sattler's layer has medium diameter blood vessels, and the choriocapillaris consists of capillaries.

Figure Description

Cross section of sclera and choroid, adapted from Wolff, 1968

Key

A episclera
B sclera
C suprachoroidia
D Haller's layer of vessels
E Sattler's layer of vessels
F choriocapillaris layer
G retinal pigment epithelium
H pigment (melanin)
I Bruch's membrane

The capillaries of the choriocapillaris are among the largest capillaries in the body, with a large lumen and thin vascular wall. The capillaries are leaky. The retinal pigment epithelium serves as a blood-retina barrier.

The choroid is thickest at the posterior pole, where metabolic demands of the retina are greatest. Average choroid thickness is approximately 250 microns at the macula, and it thins anteriorly. At the ora serrata, the choroid transitions into the supraciliaris and vascular layer of the ciliary body.

The innermost layer of the choroid is the basal lamina of the choriocapillaris, which forms the outermost layer of Bruch's membrane. The choroid is firmly attached to the retina by Bruch's membrane.

Sympathetic innervation to the choroid results in vasoconstriction. The choroid also contains intrinsic choroidal neurons, which lead to vasodilation through vasoactive intestinal peptide and nitric oxide.

THE SCLERA AND CHOROID

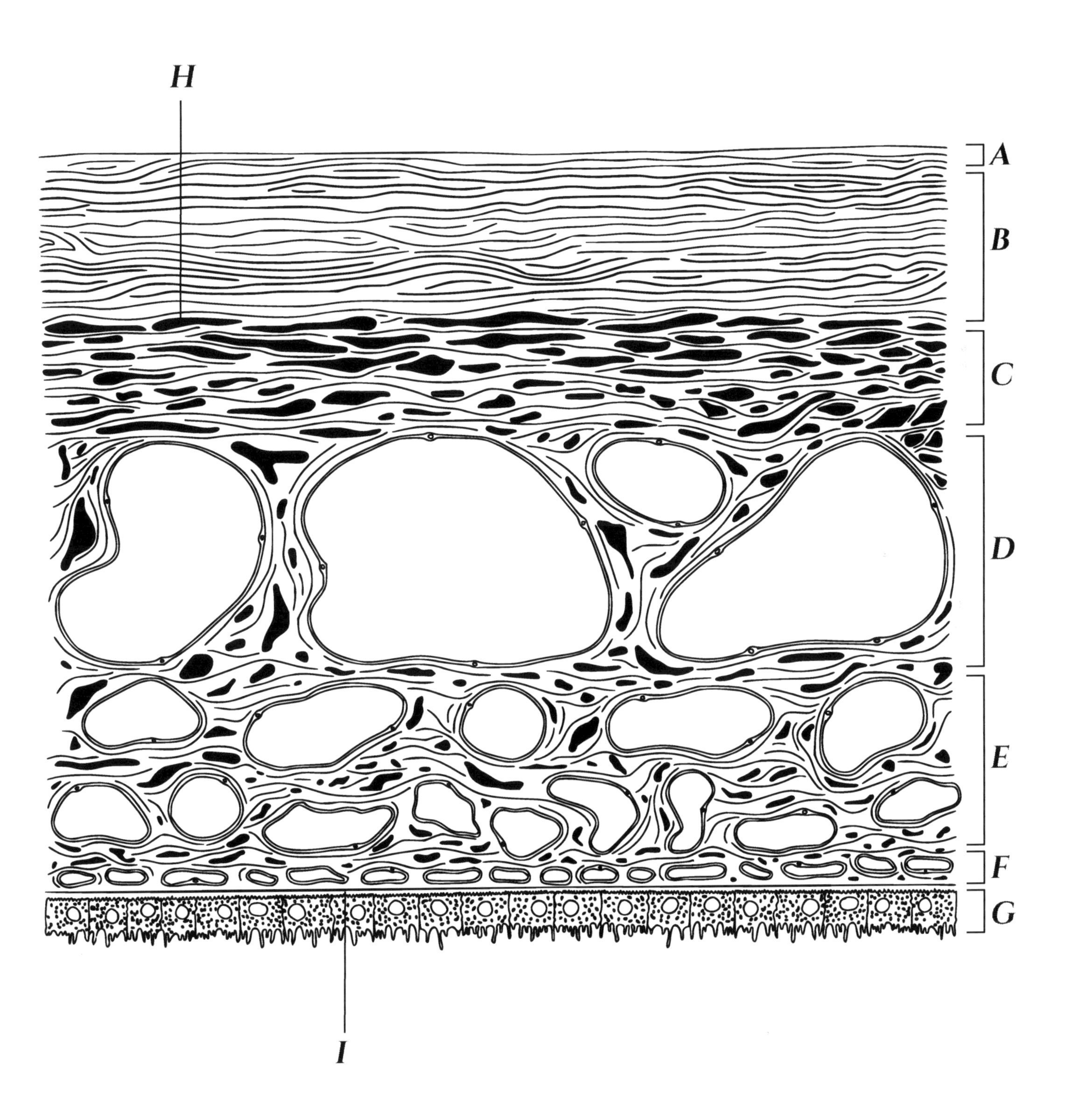

The Fundus

The fundus refers to the back surface of the eye, or posterior pole, which is visible on ophthalmoscopy. Within the fundus, the retina, optic disc, and inner retinal vasculature can be seen.

The optic disc, or optic nerve head, is the most anterior portion of the optic nerve. It appears as a prominent white area in the nasal region of the posterior pole. The disc is typically elliptical in shape, with a vertical diameter of 1.75 mm and horizontal diameter of 1.5 mm. The optic disc is the point of exit for ganglion cell axons as they form the optic nerve.

The central 5.5 mm diameter of the posterior pole is the macula, subtending a visual angle of 18°. Within the macular region, the inner segments of photoreceptors contain three types of xanthophyll pigments, zeaxanthin, meso-zeaxanthin, and lutein. These macular pigments serve to absorb short wavelength blue light, acting in an antioxidant and also having an optical role.

The macula is divided into annular regions centered around the foveal pit. The foveal pit is the thinnest portion of the retina with highest density of cone photoreceptors. The foveal pit is located about 0.5 mm inferior and 5 mm temporal to the center of the optic disc. The area just surrounding the foveal pit is the foveola, with a diameter of 0.35 mm. The foveola is within the fovea, which has a diameter of 1.5 mm. The annulus surrounding the fovea is the parafovea, which is 0.5 mm thick, and surrounding that is the perifovea, which is 1.5 mm thick.

The inner retinal vasculature is supplied by the central retinal artery and vein, which emerge from and exit the eye at the optic disc. The central retina artery is a branch of the ophthalmic artery. The vessels arch around the fovea, forming superior and inferior arcades. About 20% of the population has a cilioretinal artery, which is a branch of the short posterior ciliary arteries. The cilioretinal artery also emerges from the optic disc. It supplies the retina between the optic nerve head and fovea, and serves as a redundant vascular supply that spares the fovea in the case of central retinal artery occlusion.

Figure Description

Fundus of the left eye

Key

A nasal retina
B temporal retina
C foveal pit
D foveola
E fovea
F parafovea
G perifovea (outer limit of macula)
H optic cup
I neuroretinal rim of optic disc
J superior vascular arcade
K inferior vascular arcade
L cilioretinal artery

The Fundus

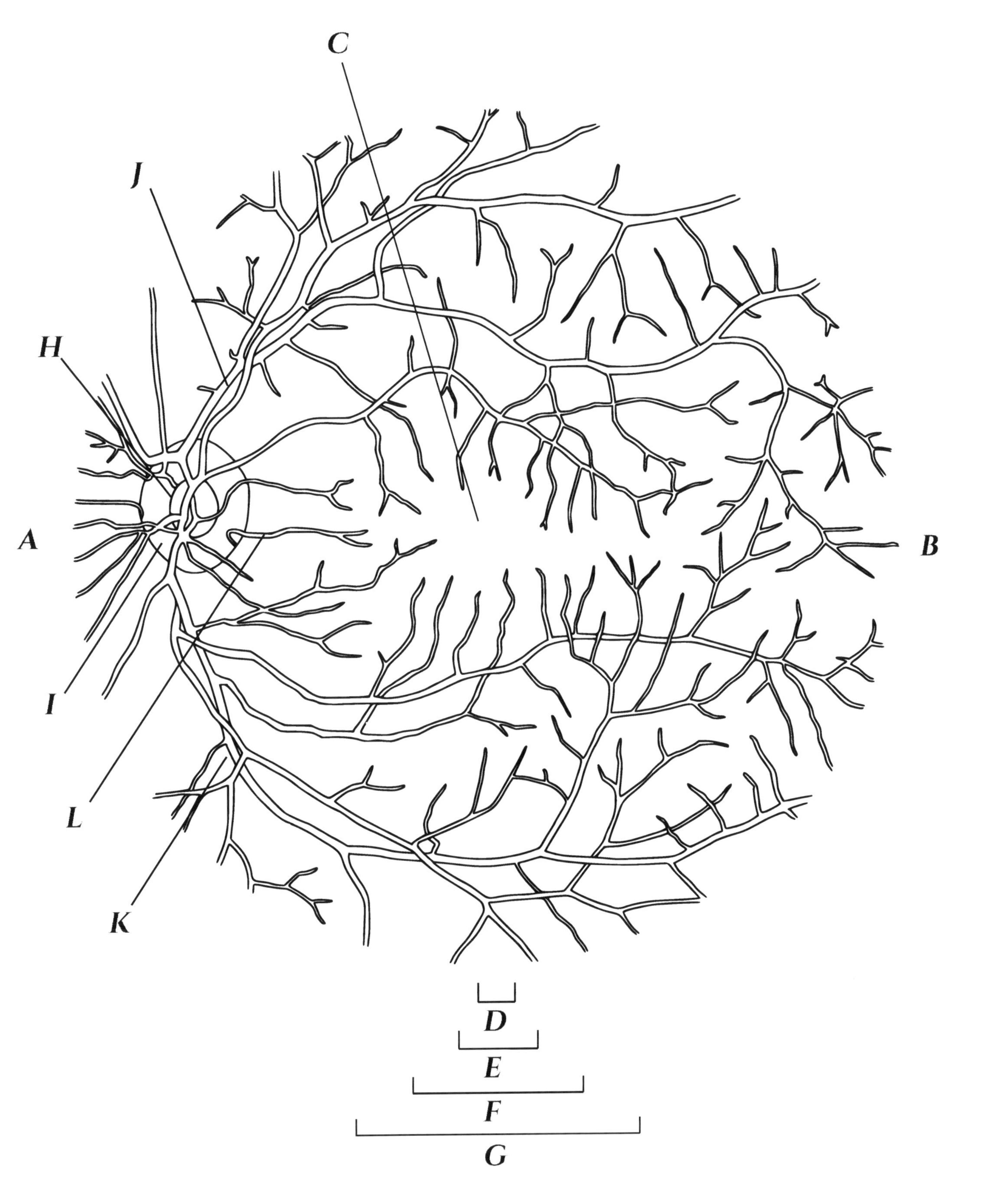

The Foveal Avascular Zone

The central avascular region of the macula is known as the foveal avascular zone. The fovea is devoid of inner retinal vasculature and is supplied solely by the choroid. The absence of inner retinal blood vessels at the fovea serves to allow light to reach the foveal cone photoreceptors unimpeded, thereby minimizing light scatter and increasing optical quality.

Figure Description

Inner retinal vasculature surrounding the fovea, adapted from Hogan, Alvarado, Weddell, 1971

The Foveal Avascular Zone

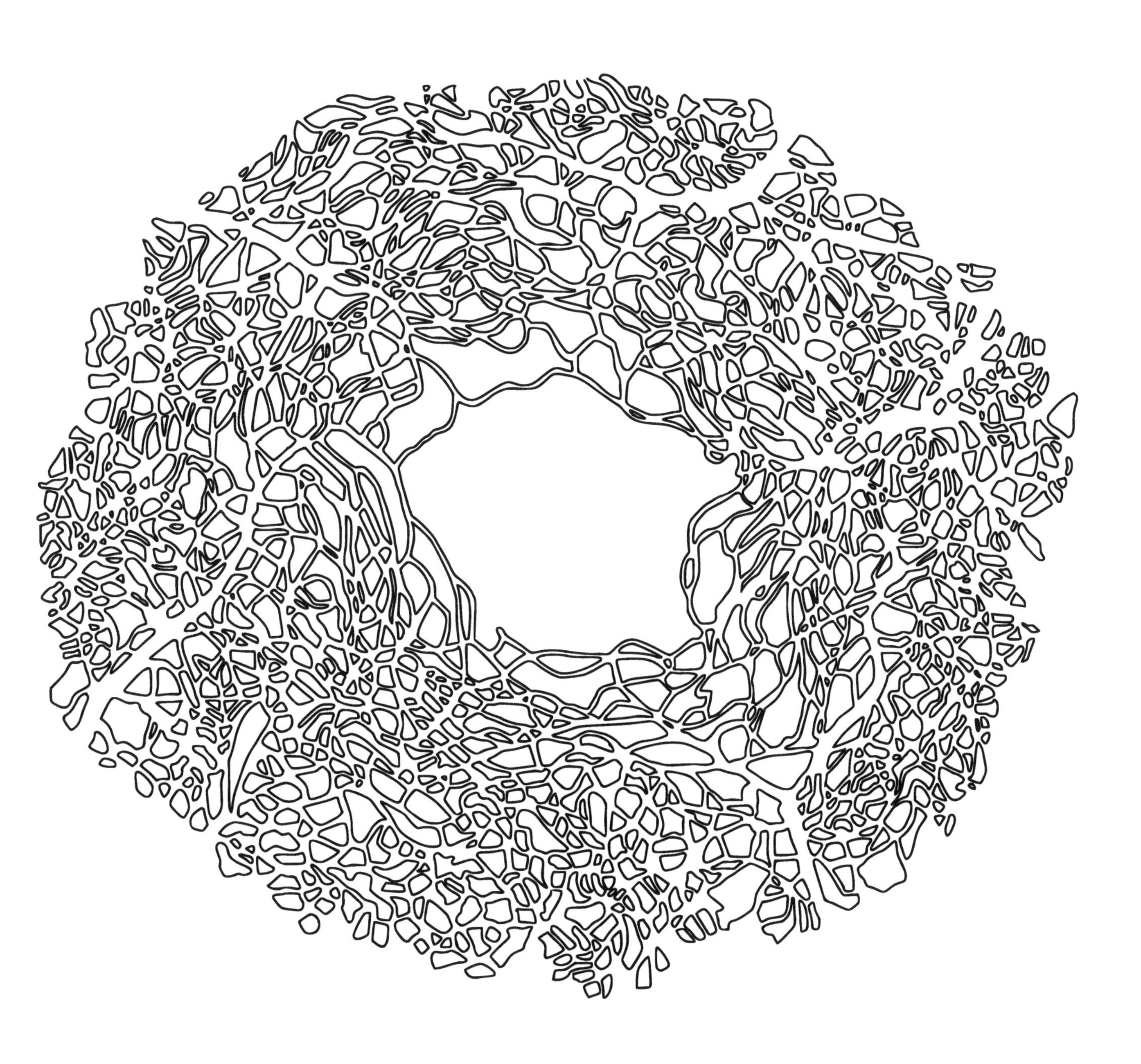

The Retina

The retina is the neurosensory component of the eye, serving to capture light energy and transmit it into a signal that is ultimately perceived as an image. The retina is derived from neural ectoderm in development. The retina is generally thought of as having ten layers, which are numbered as such, beginning with the outermost layer:

1) The retinal pigment epithelium (RPE) is a single layer of pigmented cuboidal epithelial cells between Bruch's membrane and the photoreceptors.

2) Photoreceptor outer/inner segment layer consists of the outer and inner segments of rod and cone photoreceptors. The photoreceptor outer segments are responsible for capturing photons of light. From photoreceptors, the signal will travel to bipolar and ganglion cells, with modulation by horizontal and amacrine cells.

3) The external limiting lamina consists of end processes of Müller cells. This structural layer serves to hold the photoreceptors in a very regular array. Desmosomal attachments are found between the Müller cell end processes and photoreceptor inner segments.

4) The outer nuclear layer consists of rod and cone nuclei. Cone nuclei are found in the more outer region of this layer, and rod nuclei in the more inner region.

5) The outer plexiform layer consists of synapses between photoreceptors and bipolar and horizontal cells. These are the first synapses in the visual pathway.

6) The inner nuclear layer consists of the cell bodies of bipolar cells, as well as nuclei of horizontal, amacrine, and Müller cells. Bipolar cells carry information from photoreceptors to ganglion cells. Horizontal cells provide lateral inhibition of the visual signal, inhibitory feedback to photoreceptors, and feedforward inhibition to bipolar cells. Amacrine cells carry information laterally, synapsing with bipolar and ganglion cells. Amacrine cells modulate the visual signal by providing temporal input. Müller cells are a glial cell. While Müller cells are not directly involved in transmitting the visual signal, they provide structural and physiological support to other cells. Bipolar and Müller cell nuclei tend to be found in the middle of this layer, whereas horizontal cell nuclei are in the outer region and amacrine cell nuclei are in the inner region.

Figure Description

Cross section of retina, from retinal pigment epithelium (top) to inner limiting lamina (bottom)

Key

A	retinal pigment epithelium layer
B	photoreceptor outer/inner segment layer
C	external limiting lamina
D	outer nuclear layer
E	outer plexiform layer
F	inner nuclear layer
G	inner plexiform layer
H	ganglion cell layer
I	nerve fiber layer
J	internal limiting lamina
K	retinal epithelial cell
L	rod photoreceptor
M	cone photoreceptor
N	horizontal cell
O	bipolar cell
P	Müller cell
Q	amacrine cell
R	ganglion cell

Blood vessels of the deep capillary network of the inner retinal vasculature are found in the inner nuclear layer.

7) The inner plexiform layer consists of synapses between bipolar, ganglion, and amacrine cells. The inner plexiform layer is sublaminated, with off bipolar cells synapsing in the outer region and on bipolar cells synapsing in the inner region.

8) The ganglion cell layer consists of the cell bodies of ganglion cells and displaced amacrine cells.

9) The nerve fiber layer consists of the axons of ganglion cells. Blood vessels of the superficial capillary network of the inner retinal vasculature are found in the nerve fiber layer.

10) The internal limiting lamina is the innermost layer of the retina, formed from the end processes of Müller cells. This layer creates a boundary between the retina and vitreous.

The Retina

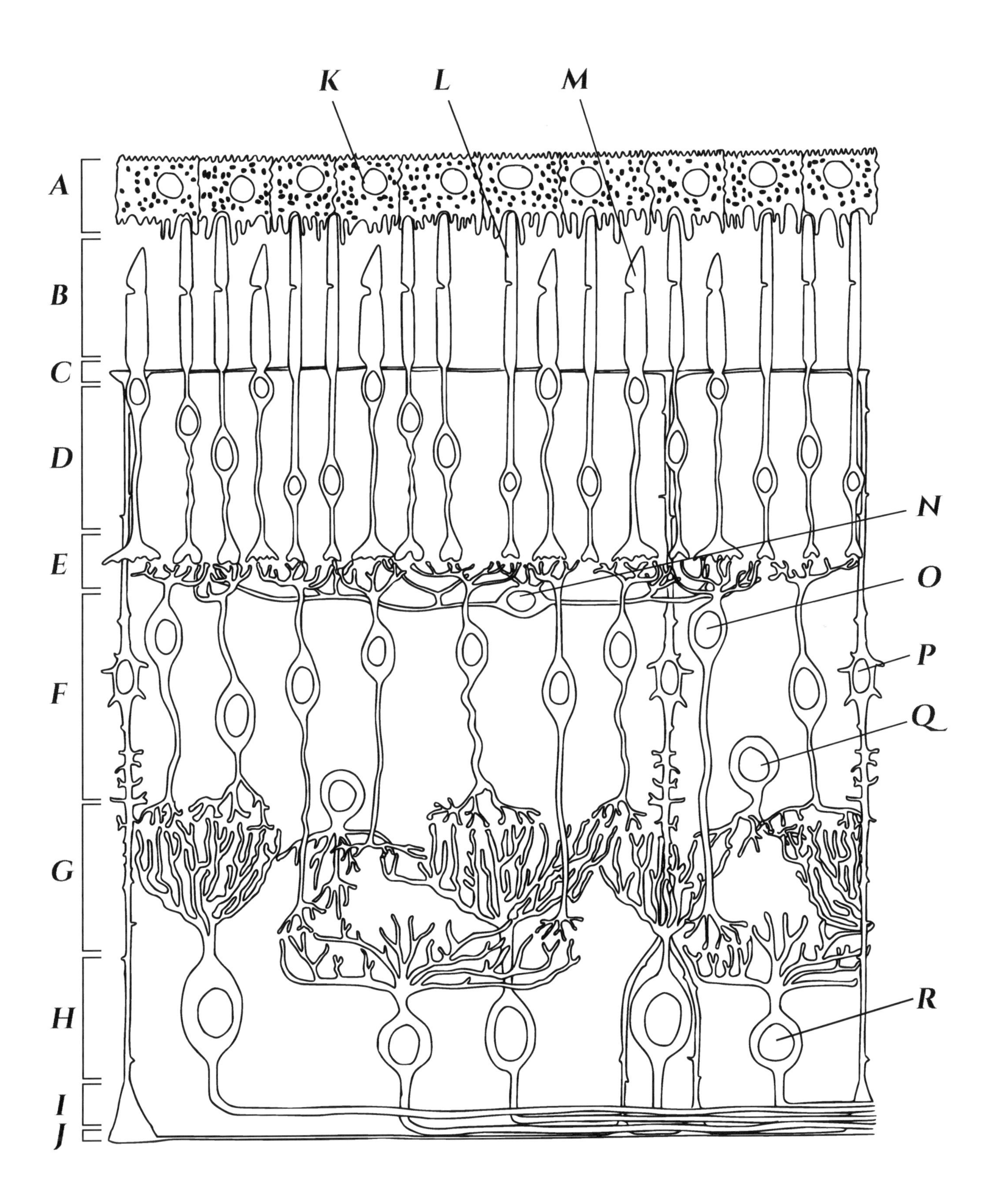

The Fovea

The fovea is a highly specialized region of the retina that is adapted for high spatial resolution. The highest density of cone photoreceptors is found at the fovea. At the foveal pit, the only retinal layers present are the retinal pigment epithelium, photoreceptor outer/inner segments, external limiting lamina, a limited number of cells in the outer nuclear layer, Henle's fiber layer, and the internal limiting lamina. There is no inner retinal vasculature at the fovea.

The cone outer and inner segments are elongated and tightly packed at the foveal pit. Other retinal layers are displaced laterally so that light can more efficiently be captured by the central cones. The central 1° of the fovea contains only red and green cones. There are no blue cones or rods in this region.

The lateral displacement of retinal layers results in the foveal pit being thinnest region of the retina, at approximately 90 µm. The nuclei and axons of the central cones travel laterally to the parafovea, creating the clivus, which is the sloped edge of the foveal pit. The diagonally traveling photoreceptor axons make up Henle's fiber layer. Moving outward from the foveal pit, the inner nuclear, outer plexiform, ganglion cell, and nerve fiber layers appear. The parafovea contains the highest number of bipolar and ganglion cells, making it the thickest region of the retina, at approximately 300 µm.

Figure Description

Cross section of retina, centered on the fovea, adapted from Wolff, 1968

Key

A retinal pigment epithelium
B photoreceptor inner/outer segment layer
C outer nuclear layer
D Henle's fiber layer
E inner limiting lamina
F inner retinal vessel
G foveal pit
H inner nuclear layer
I inner plexiform layer
J ganglion cell layer
K nerve fiber layer

THE FOVEA

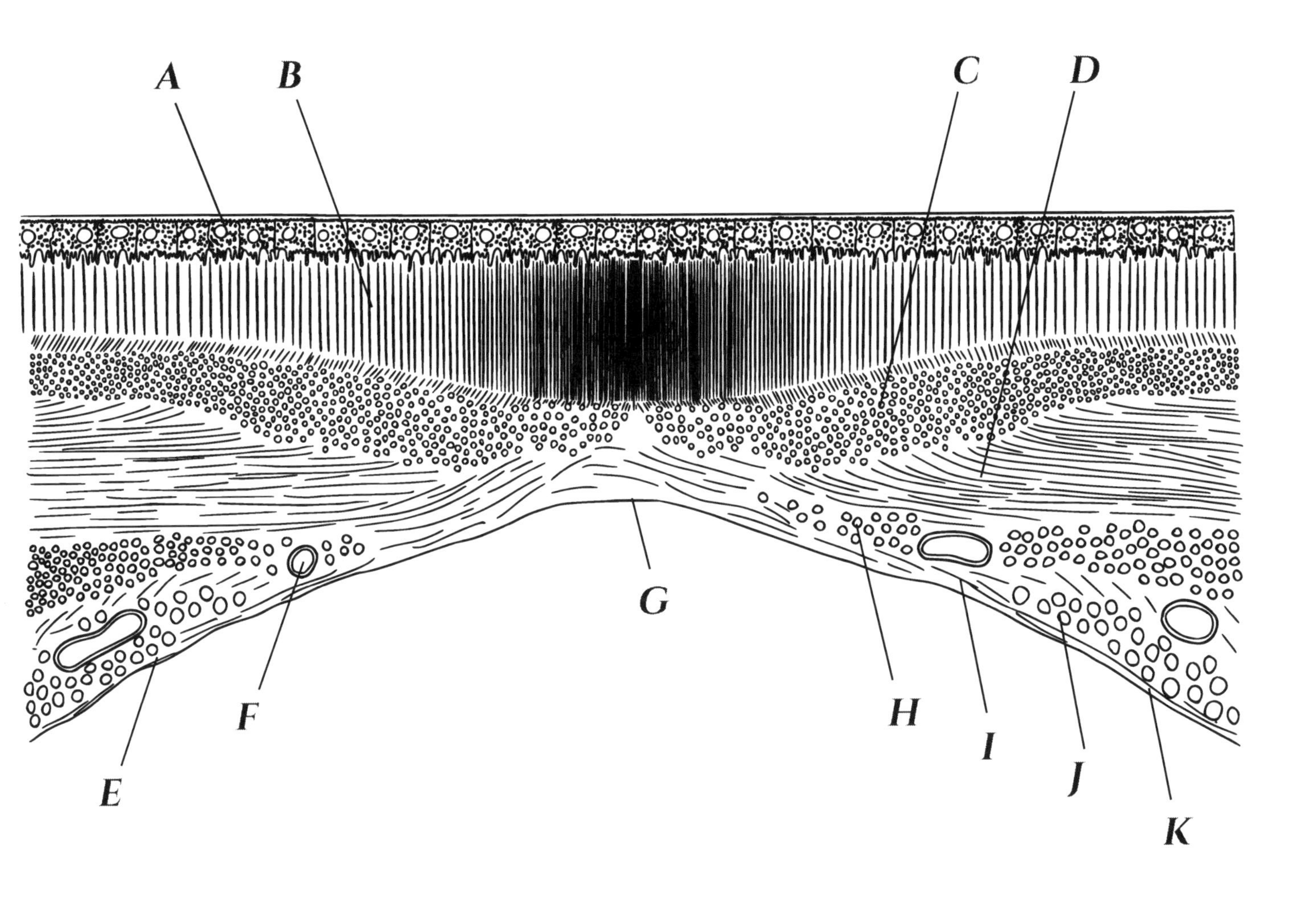

Bruch's Membrane and Retinal Pigment Epithelium

Bruch's membrane is a 2-4 µm layer between the choriocapillaris and the retinal pigment epithelium (RPE). Bruch's membrane extends from the optic disc to the ora serrata, where it is continuous with the basal lamina of the ciliary body.

Bruch's membrane functions to pass nutrients from the choroid to the retina and waste from the retina to the choroid. Bruch's membrane consists of five layers. From the outermost layer, these are the 1) basement membrane of the choriocapillaris, 2) outer collagenous layer, 3) elastic layer, 4) inner collagenous layer, and 5) basement membrane of the RPE. The inner collagenous layer is the location where drusen form, A druse is a build up of lipofuscin and other waste material. Drusen are characteristic in age-related macular degeneration.

The RPE is a single layer of hexagonal (in en face view) cuboidal epithelial cells that form the blood-retina barrier through numerous cell to cell junctions, including tight junctions, between neighboring cells. The RPE basement membrane is part of and tightly adhered to Bruch's membrane. The basal membrane of the RPE cells is convoluted. The nucleus, mitochondria, and other organelles are found towards the basal aspect of the cells. The apical aspect is highly pigmented with microvilli that extend around the outer segments of the photoreceptors. There are no junctions adhering the RPE to the photoreceptors. The space around the photoreceptor outer segments is the subretinal space.

The RPE functions to phagocytize shed photoreceptor discs, transport molecules and ions between the choroid and retina, store and metabolize vitamin A, absorb light, and produce growth factors.

Figure Description

Histological cross section of choriocapillaris (top), Bruch's membrane (middle), and retinal pigment epithelium (bottom), adapted from Hogan, Alvarado, Weddell, 1971

Key

A lumen of choriocapillaris
B red blood cell
C nucleus of vessel endothelial cell
D Bruch's membrane
E basement membrane of choriocapillaris
F outer collagenous zone
G elastic layer
H inner collagenous zone
I basement membrane of retinal pigment epithelium
J nucleus of retinal pigment epithelial cell
K melanin granule
L apical processes of retinal pigment epithelium
M subretinal space

Bruch's Membrane & Retinal Pigment Epithelium

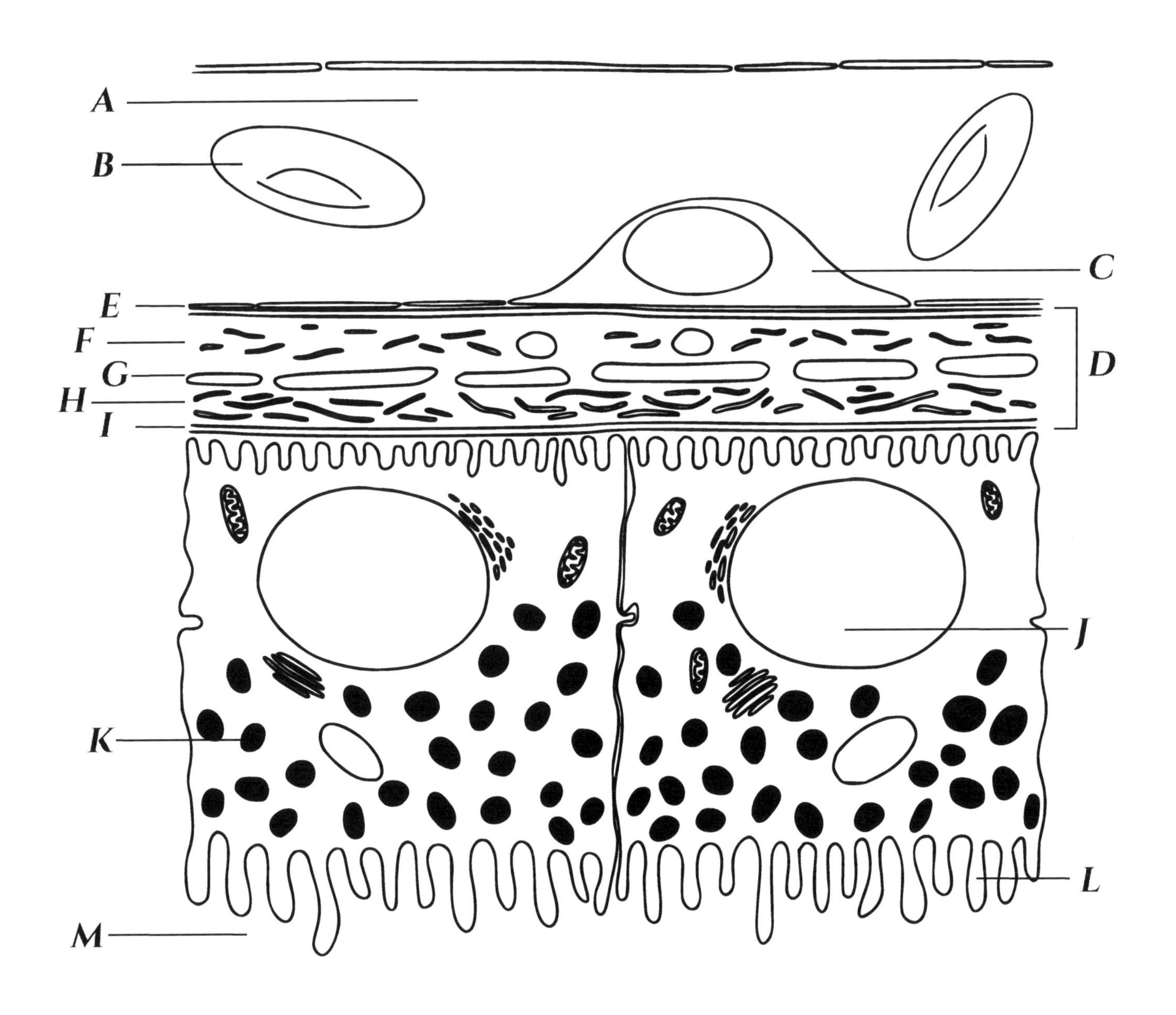

Rod and Cone Photoreceptors

Photoreceptors are the light sensitive structures of the retina. The rod and cone photoreceptors are the first-order neuron in the visual pathway, located in the outer retina. There are about 125 million rods and 6 million cones in the human retina. Rods are specialized for dim-light scotopic vision, with low spatial resolution and no color sensitivity. Cones are specialized for photopic color vision and have high spatial resolution. There are three types of cones, classified based on the wavelengths to which they are most sensitive: blue (short wavelength) cones with peak sensitivity at 440 nm, green (middle wavelength) cones with a peak sensitivity at 535 nm, and red (long wavelength) cones with a peak sensitivity at 565 nm. Cones are most numerous at the fovea, with density decreasing exponentially with eccentricity. Rods are most numerous in the mid-periphery, at approximately 20°. At the center of the fovea, there are only red and green cones, with no blue cones or rods.

The general structure of rods and cones is similar. The outer most portion of the cells is the outer segment, which consists of stacked photopigment-containing discs. Rod discs contain the photopigment rhodopsin, and cone discs contain iodopsin. Rod discs are independent of the cell membrane, whereas cone discs are formed from infoldings of the cell membrane. The outer segment is connected to the inner segment via a connecting cilium.

The inner segment contains cellular organelles and is divided into two regions, the outer ellipsoid region and inner myoid region. The ellipsoid region has a high number of mitochondria that provide metabolic energy for synthesizing new photoreceptor discs. The myoid region contains endoplasmic reticulum and golgi apparatus.

Moving inward from the inner segment, the cells pass through the external limiting lamina towards the nuclei in the outer nuclear layer. The cone nucleus is close to the cone inner segment, whereas the rod nucleus is farther away, such that there is a rod outer fiber leading to the nucleus. From the nucleus, an inner fiber (the axon of the cell) leads to the synaptic terminal in the outer plexiform layer.

The synaptic terminal of the rod is a spherule, and of a cone is a pedicle. Rod spherules contain one invaginating ribbon synapse, termed a triad synapse; there are dendrites of two horizontal cells and one bipolar cell in a triad synapse.

The cone pedicle is much larger than the rod spherule and consists of up to 30 invaginating ribbon synapses and numerous flat synapses. Similar to the rod spherule, the dendrites of two horizontal cells and one bipolar cell are found in the invaginating synapse. The flat synapses of the cone pedicle consist of synapses between the cone and OFF-bipolar cells.

In the pre-synaptic terminal, glutamate-containing vesicles collect around a ribbon-like structure; hence, the name ribbon synapse. Rods and cones release glutamate in the dark. When light is captured, the visual cascade is initiated. The cell hyperpolarizes and glutamate release stops.

Figure Description

A rod (left) and cone (right) photoreceptor

Key

A rod outer segment
B rod inner segment
C rod ellipsoid region
D rod myoid region
E cone outer segment
F cone inner segment
G cone ellipsoid region
H cone myoid region
I photoreceptor discs
J connecting cilium
K mitochondria
L endoplasmic reticulum and golgi apparatus
M rod outer fiber
N nucleus
O inner fiber (i.e. axon)
P rod spherule
Q cone pedicle
R ribbon synapse

Rod & Cone Photoreceptors

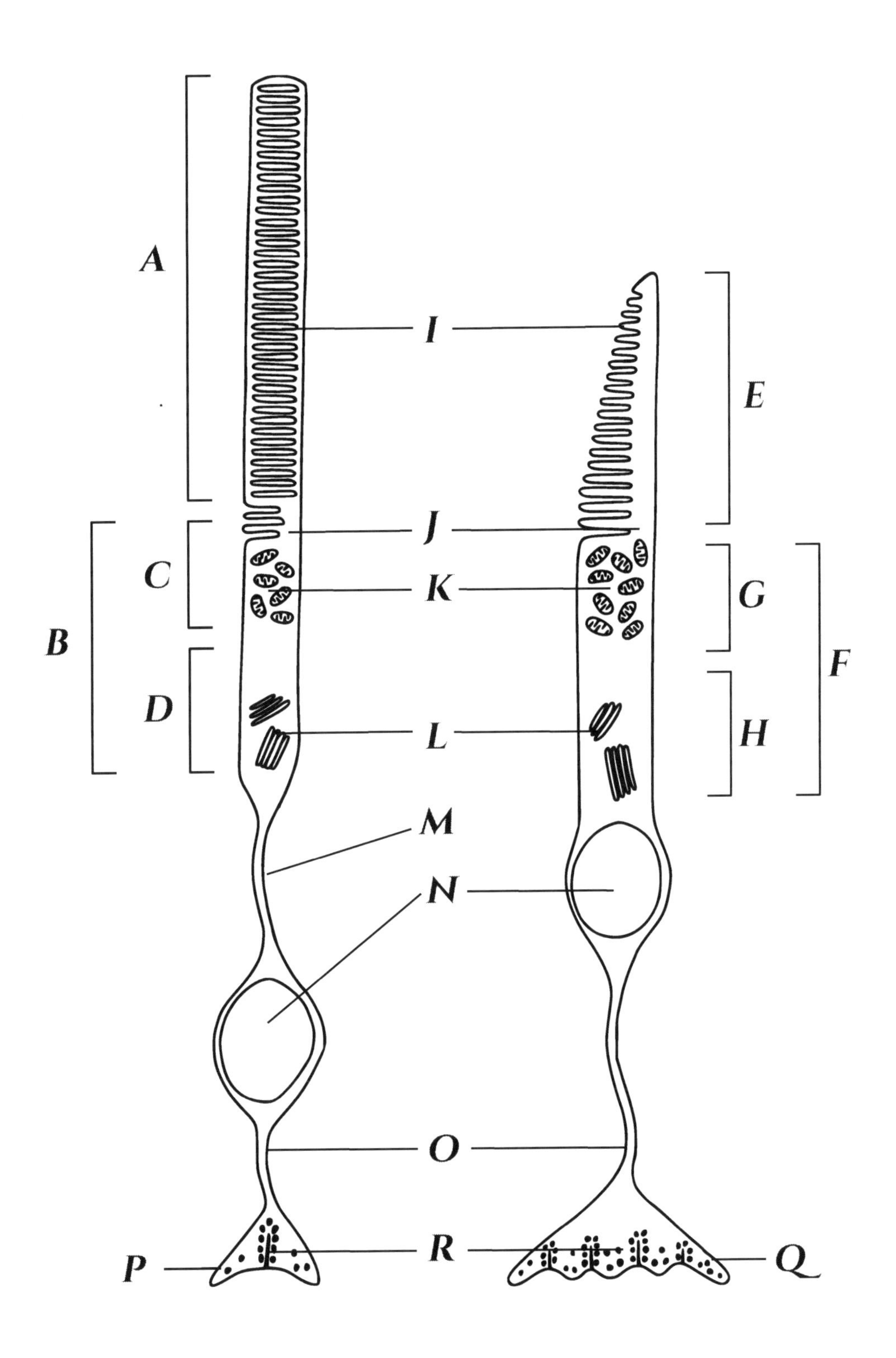

Photoreceptor Disc

Photoreceptor discs are photopigment-containing organelles in the outer segment of rods and cones. In rods, discs are independent of the cell membrane, resembling a stack of pita bread in the outer segment. In cones, discs are continuous with the cell membrane. Disc membranes are formed from a phospholipid bilayer.

Photopigment opsins are embedded in the disc membrane and absorb light. Rod discs contain rhodopsin, and cones contain wavelength-specific opsins. Opsins are a group of proteins that are made light sensitive via retinal, a vitamin A-related chromophore. Human visual opsins are seven transmembrane domain proteins that serve as g-protein coupled receptors in the rods and cones. They are involved in the conversion of light to neural signals in the photoreceptors.

The photopigment is embedded in the disc membrane. Each photoreceptor has approximately 1000 discs, and each disc is embedded with approximately 1000 photopigment molecules.

Other molecules embedded in the disc membrane include ABCA4 (a member of the ATP-binding cassette transporter gene), adenosine triphosphate (an organic compound and hydrotrope that provides energy), photoreceptor disc component (a protein coding gene), peripherin (a type III intermediate filament protein), and other proteins (not shown).

Photopigment is formed in the ellipsoid region of the photoreceptor inner segment and assembled into the discs near the cilium. Discs undergo continuous shedding and renewal. Discs migrate along the outer segment towards the RPE, where they are shed in a circadian manner. Rod discs shed in the morning at light onset, and cone discs shed in the evening at light offset.

Figure Description

A rod photoreceptor outer segment disc, adapted from Skiba et al. 2013

Key

A phospholipid bilayer
B rhodopsin molecule

Rod Photoreceptor Disc

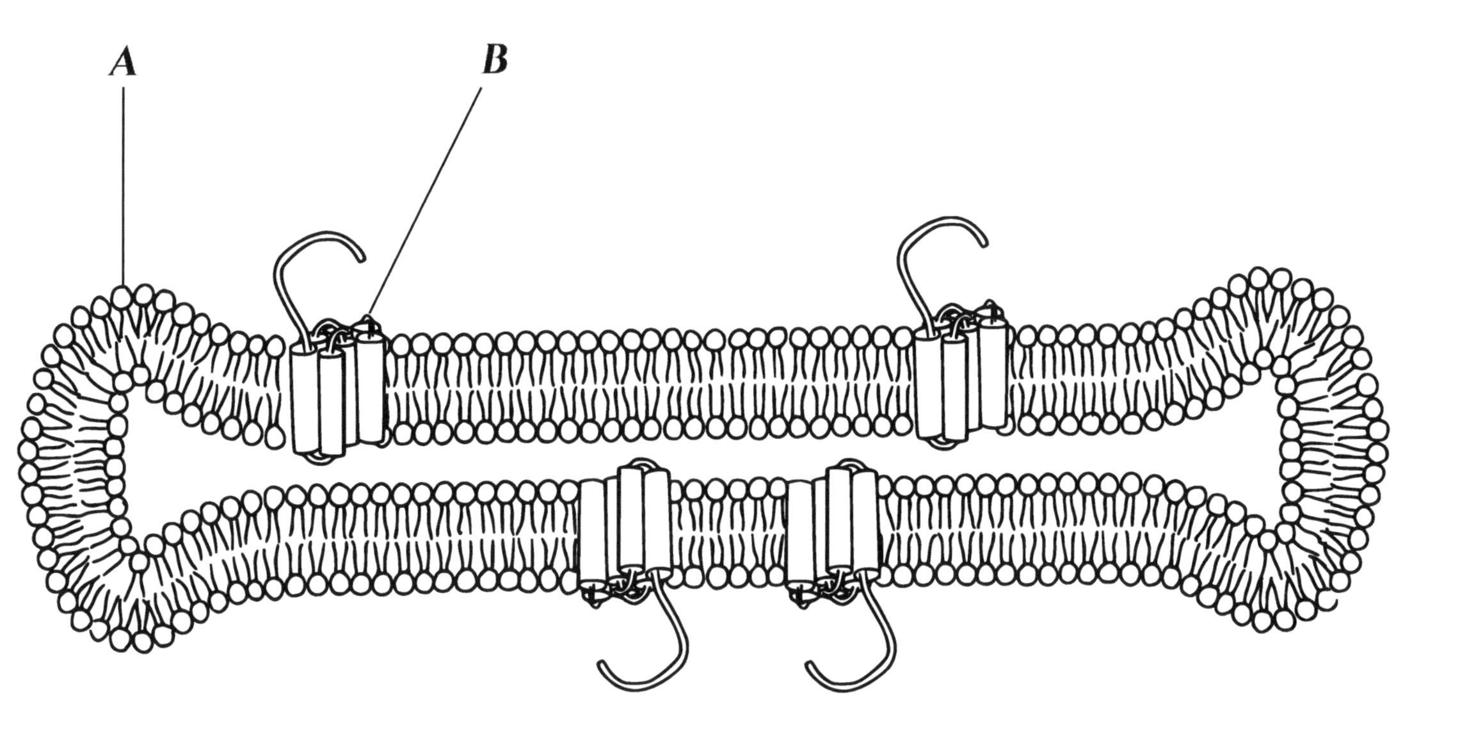

Bipolar Cells

Bipolar cells are the second order retinal neuron in the visual pathway. Bipolar cells convey the visual signal from photoreceptors to ganglion cells. Their dendrites are found in the outer plexiform layer, nuclei are in the inner nuclear layer, and terminal synapses are in the inner plexiform layer. Near the fovea, there is a one to one correspondence between photoreceptors and midget bipolar cells. Elsewhere, numerous photoreceptors may converge onto single diffuse bipolar cells, contributing to a convergence of information.

There are several types of bipolar cells with varying function and morphology. Bipolar cells can be either "ON" or "OFF" subtypes. ON bipolar cells are stimulated when the light is turned on and synapse in the more inner ON sublamina of the inner plexiform layer. OFF bipolar cells are stimulated when the light is turned off and synapse in the more external OFF sublamina of the inner plexiform layer.

Rod photoreceptors synapse with only one type of rod bipolar cell, an ON bipolar cell. One rod bipolar cell may contact up to 40 rod spherules. Rod bipolar cells do not directly synapse with ganglion cells, but rather, join the cone bipolar cell circuitry to reach the ganglion cells.

On the other hand, there are at least 10 sub-classes of cone bipolar cells. ON cone bipolar cells synapse at the invaginating ribbon synapses of the cone pedicle. OFF cone bipolar cells synapse at the flat surface of the cone pedicle. At the fovea, midget bipolar cells connect one cone to one retinal ganglion cell. The one-to-one communication allows for increased spatial resolution. Diffuse bipolar cells synapse with 5-10 cones.

Figure Description

Retinal bipolar cells, adapted from Kolb

Key

A blue cone bipolar cell
B diffuse OFF bipolar cell
C diffuse ON bipolar cell
D midget OFF bipolar cell
E rod bipolar cell
F dendrites
G nuclei
H synaptic terminals in OFF sublamina
I synaptic terminals in ON sublamina

BIPOLAR CELLS

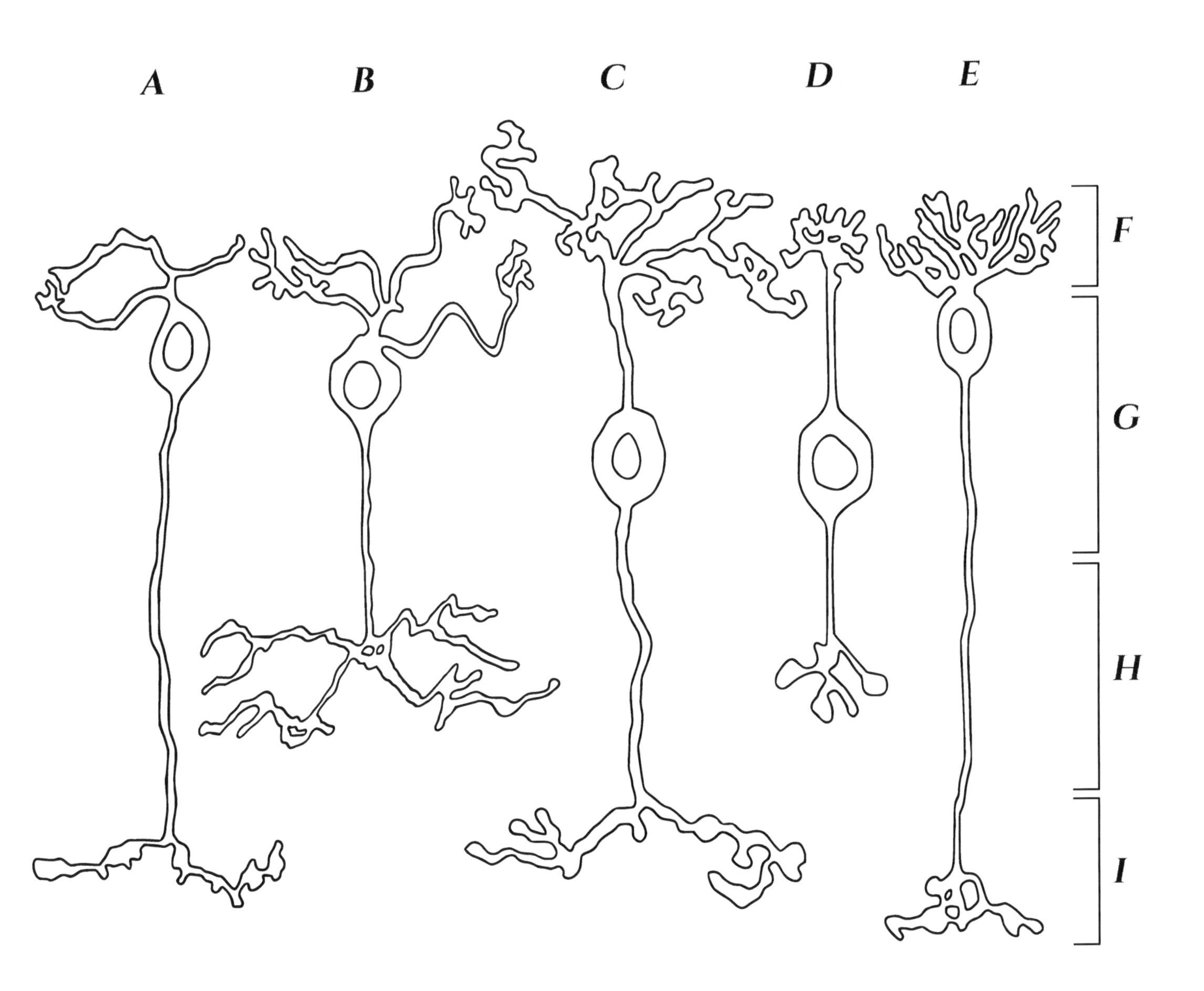

Retinal Ganglion Cells

Retinal ganglion cells (RGCs) are the third order neuron in the visual pathway. Their dendrites synapse with bipolar and amacrine cells in the inner plexiform layer, their nuclei are in the ganglion cell layer, and their axons make up the nerve fiber layer. There are approximately 750,000 to 1.5 million RGCs in the human retina.

There are over 20 different types of RGCs. The majority of RGCs project to the lateral geniculate nucleus (LGN) of the thalamus and are involved in image formation. Some RGCs project to the superior colliculus and function in eye movement. Other RGCs, known as the intrinsically photosensitive RGCs (ipRGCs), project to the suprachiasmatic nucleus, functioning in circadian rhythm entrainment, and to the olivary pretectal nucleus, functioning in pupil size regulation.

RGCs are classified based on their dendritic arbor morphology and projections to the LGN. Parvocellular (P) RGCs, aka midget cells, project to the parvocellular layers of the LGN. P cells are the most common RGC, containing one dendritic synapse with one midget bipolar cell, which synapses with one foveal cone. P cells have a small diameter axon. They have high spatial resolution, are involved in color vision, and have low contrast gain.

Magnocellular (M) RGCs, aka parasol cells, project to the magnocellular layers of the LGN. M cells are largely driven by rods and have large diameter axons. M cells have lower spatial resolution, are achromatic luminance cells, and have high contrast gain.

Small bistratified RGCs project to the koniocellular layers of the LGN. These cells have moderate spatial resolution, moderate conduction velocity, and can respond to moderate-contrast stimuli. These cells are involved in blue-yellow color vision.

ipRGCs make up approximately 0.4-1.5% of the ganglion cell population in the human retina. ipRGCs are intrinsically photosensitive via the photopigment melanopsin; hence, they are a photoreceptor in the inner retina. While few in number, the ipRGCs have a large dendritic spread, providing coverage across the entire retina (except directly at the fovea). ipRGCs are considered environmental illuminance detectors and convey information about night and day to the body.

Figure Description

Retinal ganglion cells, dendritic arbors, adapted from Liao et al. 2016

Key

A small bistratified retinal ganglion cell
B parasol retinal ganglion cell
C midget retinal ganglion cell
D intrinsically photosensitive retinal ganglion cell

Retinal Ganglion Cells

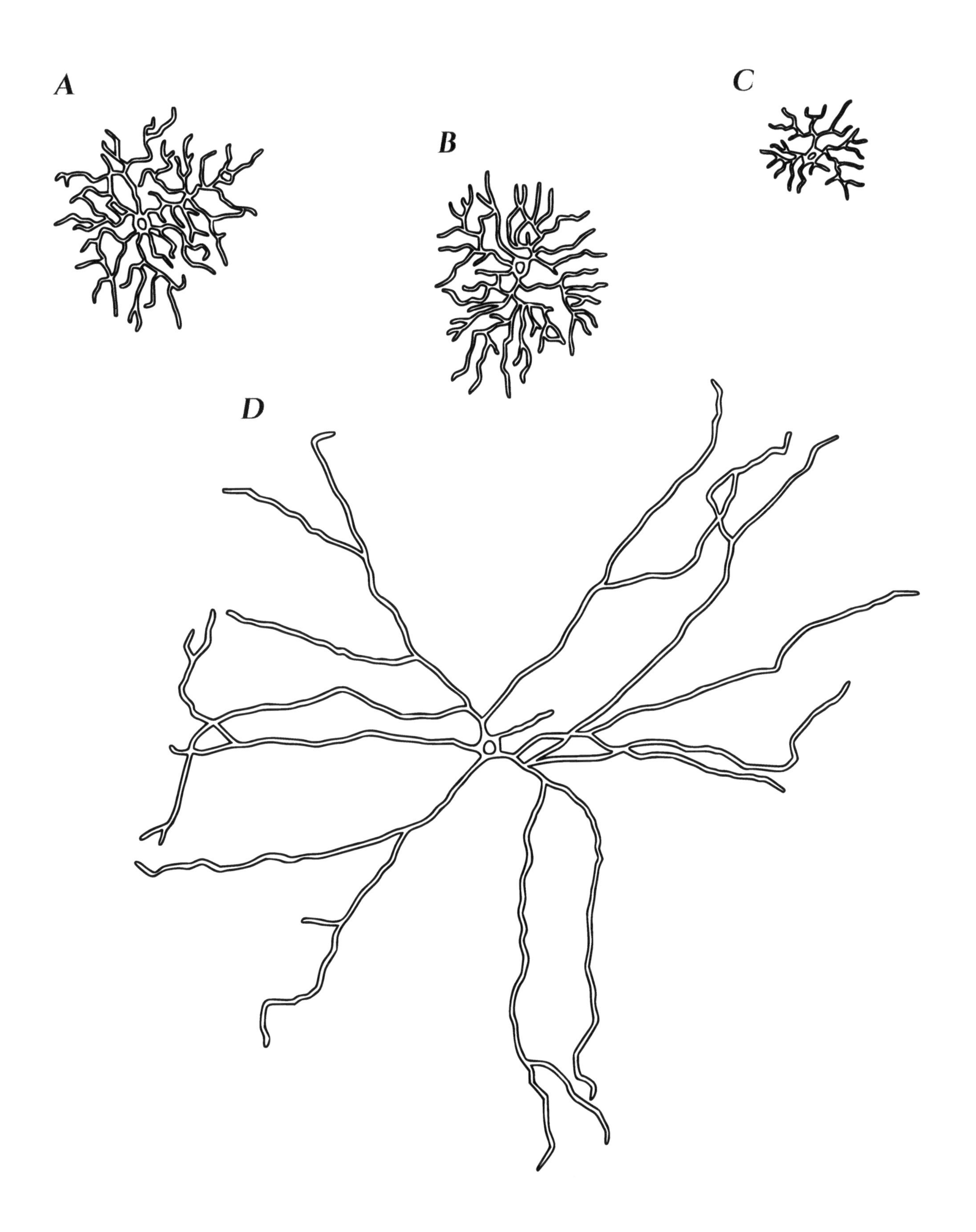

Retinal Nerve Fiber Layer

The retinal nerve fiber layer is formed by the unmyelinated axons of the retinal ganglion cells (RGCs), as the axons travel from cell bodies in the inner retina to the optic nerve head. The topography of the nerve fiber layer follows a characteristic pattern based on the distribution of RGCs in the retina. The fibers travel in such a way to circumvent the fovea. The nerve fiber layer spans of the entire retina except the fovea. It is the only retinal layer present at the optic nerve head.

Axons of RGCs in the superior retina follow the superior arcades, and in the inferior retina follow the inferior arcades, with no crossing of axons over the horizontal midline. A horizontal raphe spans from the macula temporally where the superior and inferior fibers meet.

Axons from RGCs between the fovea and optic nerve head travel in the papillomacular bundle. Axons of RGCs that are nasal to the optic nerve head travel in the nasal radial bundle.

Axons from RGCs that are more distal from the optic nerve head travel in more inner layers within the nerve fiber layer than from more proximal RGCs.

Due to the high density of ganglion cells in the parafoveal region, the nerve fiber layer is thickest around the fovea. The nerve fiber layer also thickens around the optic nerve head, where all axons converge to exit the eye as the optic nerve. Around the optic disc, nerve fiber layer thickness in healthy eyes ranges from 46 to 106 µm. It is thickest superiorly and inferiorly and thinnest medially and temporally.

Figure Description

Nerve fiber layer of the left eye

Key

A nasal retina
B temporal retina
C optic disc
D fovea
E superior arcuate fibers
F inferior arcuate fibers
G horizontal raphe
H papillomacular bundle
I nasal radial bundle

Retinal Nerve Fiber Layer

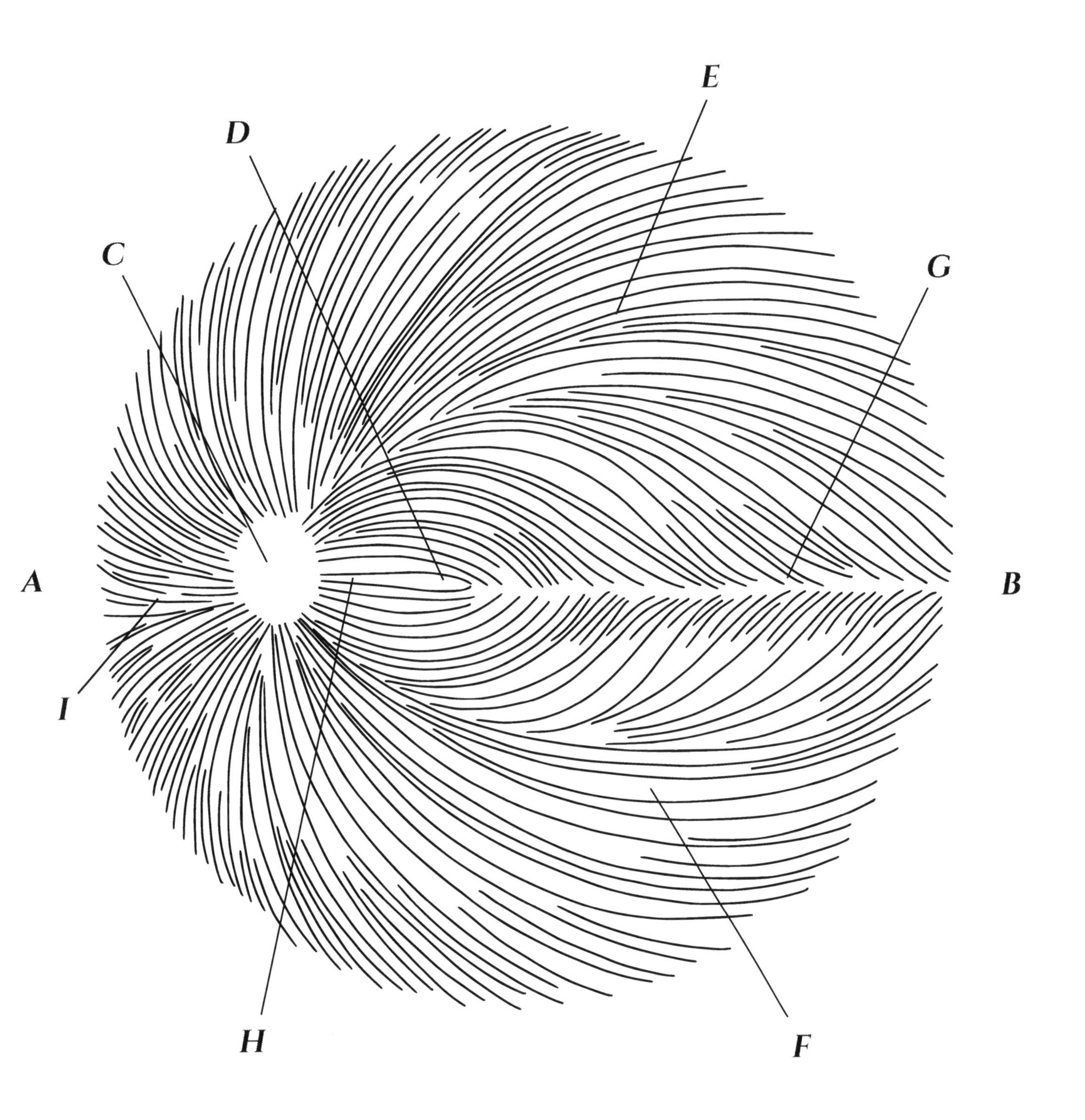

The Optic Nerve Head

The optic nerve head is the point of exit of retinal ganglion cell (RGC) axons. The axons pass through the lamina cribrosa of the optic nerve head and form the optic nerve. The optic nerve head is also the entry and exit point for the central retinal vasculature. The optic nerve head is located in the nasal region of the posterior pole.

At the optic nerve head, all of the retinal layers come to an end except the nerve fiber layer. The choroid also terminates at the optic nerve. Because there are no photoreceptors at the optic nerve head, there is a blind spot in each eye's temporal visual field.

The sclera is continuous across the optic nerve head as the collagenous lamina cribrosa, which is a sieve-like structure that the RGC axons pass through as they exit the eye. The region of the optic nerve head anterior to the lamina cribrosa is the prelaminar optic nerve. The tissue within the lamina cribrosa makes up the laminar optic nerve, and the tissue posterior to the lamina cribrosa makes up the retrolaminar optic nerve.

As RGC axons reach the optic nerve head, they make a 90° turn to exit the eye, where they become organized into nerve fascicles that are separated by glial septa. Posterior to the lamina cribrosa, the nerve fascicles are separated by pial septa, and axons become myelinated. Numerous oligodendrocytes, which produce the myelination, are found within the nerve fascicles in the retrolaminar optic nerve. The diameter of the optic nerve increases from 1.5 mm in the laminar region to 3-4 mm in the retrolaminar region.

The optic nerve is surrounded by astrocytes. At the anterior most surface of the retina, the inner limiting lamina continues at the optic nerve head; however, it is formed from astrocytes rather than Müller cells, making up the inner limiting membrane of Elschnig. At the center of the optic cup, the collection of astrocytes is the central meniscus of Kuhnt. In the prelaminar region, astrocytes surrounding the optic nerve, separating the axons from the retina, make up the intermediary tissue of Kuhnt. Moving posteriorly, the astrocytes surrounding the optic nerve in the laminar region make up the border tissue of Jacoby, and in the retrolaminar region, they make up the peripheral mantle of Graefe. The extension of sclera between the choroid and optic nerve is the border tissue of Elschnig.

Figure Description

Longitudinally oriented histological cross section of the optic nerve

Key

A retina
B choroid
C sclera
D central retinal artery
E central retinal vein
F pre-laminar nerve fascicle
G lamina cribrosa
H retro-laminar nerve fascicle
I pia mater
J arachnoid mater/space
K dura mater
L glial septum
M pial septum
N pial vasculature
O posterior ciliary artery
P central retinal vasculature
Q circle of Zinn-Haller
R inner limiting membrane of retina
S inner limiting membrane of Elschnig
T central meniscus of Kuhnt
U intermediary tissue of Kuhnt
V border tissue of Jacoby
W peripheral mantle of Graefe
X border tissue of Elschnig
Y neuroretinal rim
Z optic cup

Similar to the brain and spinal cord, the optic nerve is surrounded by three layers of meninges, including pia mater, the delicate layer of meninges adhered to the nerve tissue, arachnoid mater, the web-like middle layer of meninges, and dura mater, the tough outer layer of meninges.

The optic nerve has a rich vascular supply, originating from the ophthalmic artery. The ophthalmic artery gives off the central retinal artery, which pierces the optic nerve, traveling anteriorly and emerging from the optic nerve head to supply the inner retina. The short posterior ciliary arteries travel outside of the optic nerve, piercing the sclera and contributing to the choroid, as well as the providing branches that supply the optic nerve as pial vessels, and branches that contribute to the circle of Zinn-Haller, an arterial circle surrounding the optic nerve.

THE OPTIC NERVE HEAD

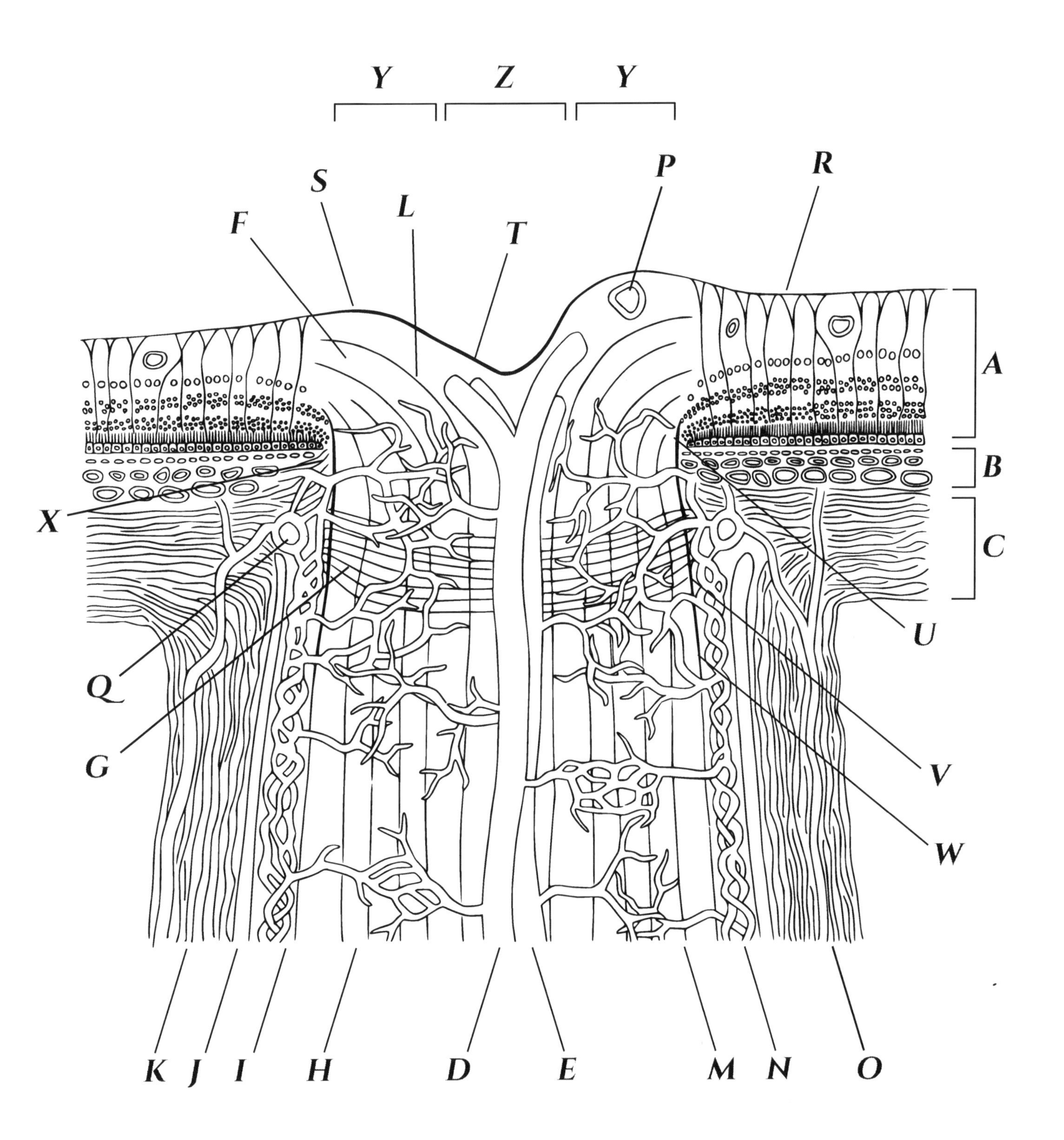

The Lamina Cribrosa

The lamina cribrosa is a sieve-like structure that spans across the scleral opening at the optic nerve head. The lamina cribrosa is composed of a multilayered network of collagenous beams which form pores that ganglion cell axons pass through as they exit the eye to form the optic nerve. The central retinal artery and vein pass through the center of the lamina cribrosa.

The lamina cribrosa functions to form a boundary between the intraocular space and the extraocular space, thereby creating a pressure gradient between intraocular pressure and intracranial pressure.

Figure Description

En face image of the collagenous tissue of the lamina cribrosa of the optic nerve head

Key

A collagenous beam
B pore
C opening for central retinal artery
D opening for central retinal vein

The Lamina Cribrosa

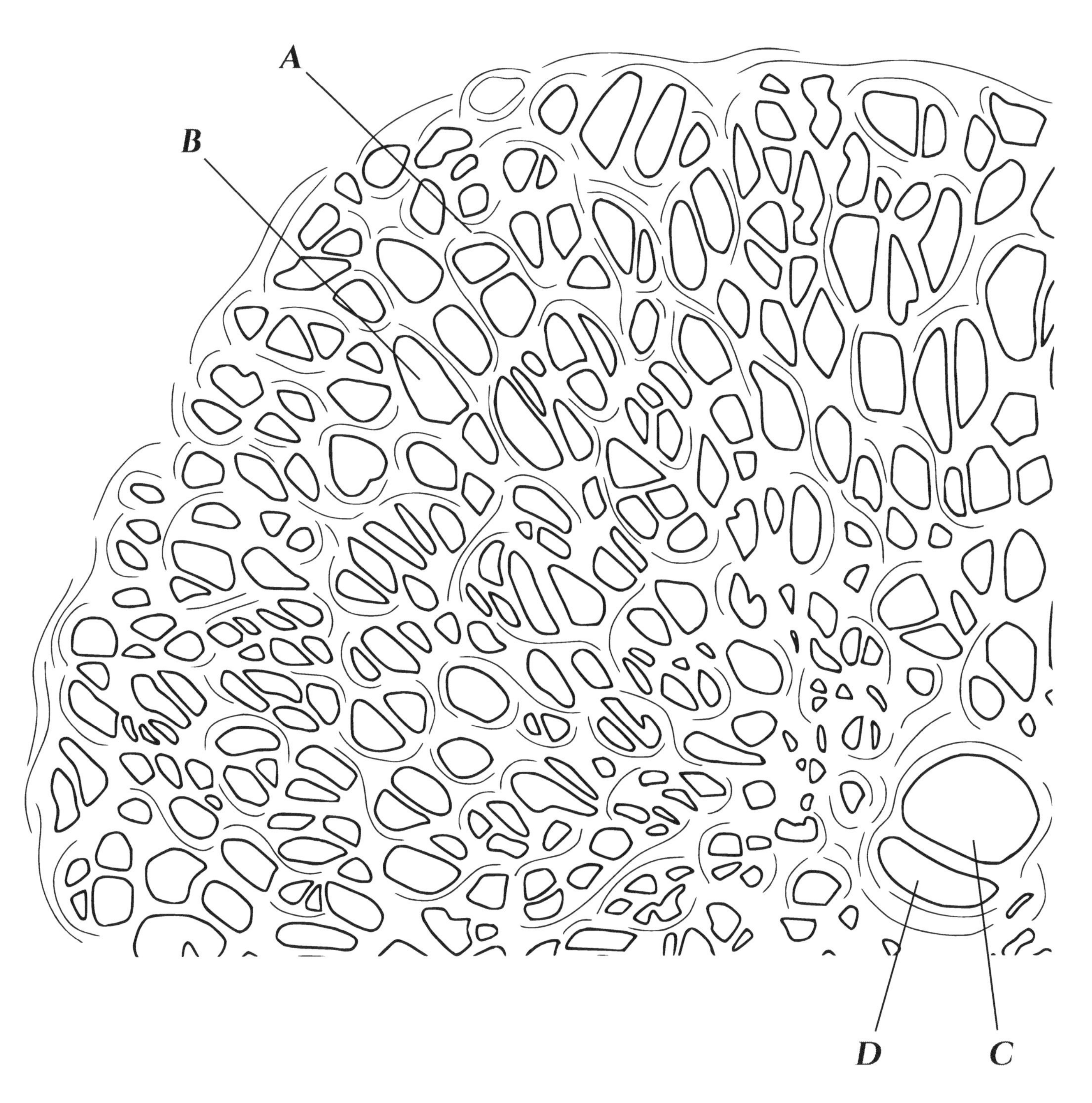

NEUROANATOMY

Cranial Nerves

The 12 paired cranial nerves emerge from the forebrain and brainstem. The cranial nerves are described using roman numerals, based on the number in which they exit the brain, from rostral to caudal, and named based on structure or function. Cranial nerves can carry motor, sensory, and/or parasympathetic information. Here, each nerve is briefly described, and in the following pages, those that are directly involved in functions related to the eye and orbit are more thoroughly discussed.

Cranial nerve I, olfactory - functions in smell, travels from the neuroepithelium of the upper nasal cavity through the cribriform plate to the olfactory bulbs, carries sensory information

Cranial nerve II, optic - functions in vision, originates in the diencephalon, carries sensory information

Cranial nerve III, oculomotor - functions in eye and eyelid movement, accommodation, and miosis, originates in the mesencephalon (midbrain), carries motor and parasympathetic information

Cranial nerve IV, trochlear - functions in eye movement, originates in the mesencephalon, carries motor information

Cranial nerve V, trigeminal - functions in sensation from the face and in mastication, originates in the metencephalon (pons), carries sensory and motor information

Cranial nerve VI, abducens - functions in eye movement, originates in the metencephalon, carries motor information

Cranial nerve VII, facial - functions in facial expression, taste, and gland secretion, originates in the metencephalon, carries motor, sensory, and parasympathetic information

Cranial nerve VIII, vestibulocochlear - functions in equilibrium and hearing, originates in the myelencephalon (medulla), carries sensory information

Cranial nerve IX, glossopharyngeal - functions in swallowing, taste, and gland secretion, originates in the myelencephalon, carries motor, sensory, and parasympathetic information

Figure Description

Ventral view of brain, highlighting the cranial nerves

Key

A	olfactory bulbs and tract
B	cranial nerve II, optic
C	cranial nerve III, oculomotor
D	cranial nerve IV, trochlear
E	cranial nerve V, trigeminal
F	cranial nerve VI, abducens
G	cranial nerve VII, facial
H	cranial nerve VIII, vestibular
I	cranial nerve IX, glossopharyngeal
J	cranial nerve X, vagus
K	cranial nerve XI, spinal accessory
L	cranial nerve XII, hypoglossal
M	cervical spinal nerve 1
N	cervical spinal nerve 2
O	optic chiasm
P	mammillary bodies
Q	cerebellum
R	pons
S	medulla
T	medullary pyramid
U	spinal cord

Cranial nerve X, vagus - functions in regulation of internal organs and gland secretion, originates in the myelencephalon, carries parasympathetic, motor, and sensory information

Cranial nerve XI, spinal accessory - functions in neck movement, originates in the myelencephalon and spinal cord, carries motor information

Cranial nerve XII, hypoglossal - functions in tongue movement, originates in the myelencephalon, carries motor information

Cranial Nerves

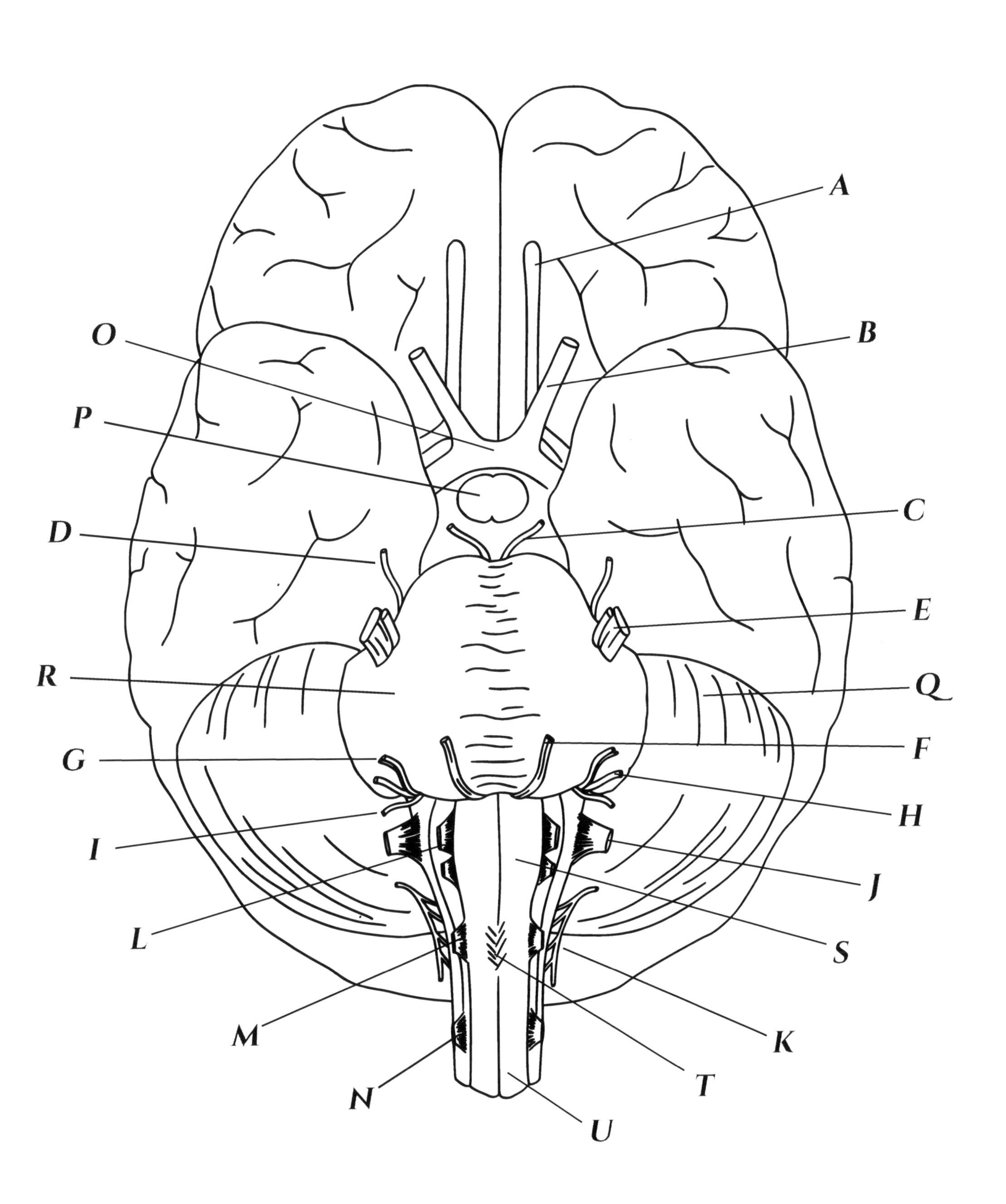

Cranial Nerve II

Cranial nerve II is the optic nerve, a special sensory afferent nerve involved in vision. While all cranial nerves are part of the peripheral nervous system, CN II has many characteristics that are central nervous system-like. For example, CN II originates from the diencephalon, rather than the brainstem. Also, CN II is ensheathed in three layers meninges, pia, arachnoid, and dura mater. Myelination of the optic nerve is produced by oligodendrocytes.

CN II carries information from retinal ganglion cells (RGCs) to the lateral geniculate nucleus (LGN) of the thalamus, as well as to the suprachiasmatic nucleus, superior colliculus, and olivary pretectal nucleus. CN II forms as axons of the RGCs exit the globe. The optic nerve travels through the optic canal in the sphenoid bone, then passes superior to the pituitary gland in the middle cranial fossa, where nerves from each eye unite at the optic chiasm. Here, fibers from the nasal retina of each eye decussate, such that nerve fibers from each left retina (right visual field) synapse in the left LGN, and from each right retina (left visual field) synapse in the right LGN. Posterior to the optic chiasm, the nerve fibers are referred to as the optic tract.

Approximately 90% of the nerve fibers in the optic tract synapse in the lateral geniculate nucleus of the thalamus, where information regarding the visual signal is then conveyed to the primary visual cortex.

Some nerve fibers synapse in the superior colliculus of the midbrain. Input is primarily from rod photoreceptors. The superior colliculus serves to coordinate eye movements, specifically, saccades and fixation.

A small number of fibers from intrinsically photosensitive retinal ganglion cells project to the suprachiasmatic nucleus and to the olivary pretectal nucleus. Fibers to the suprachiasmatic nucleus in the hypothalamus mediate circadian rhythm by conveying environmental illuminance information. Fibers that synapse in the olivary pretectal nucleus of the midbrain are involved in pupil function. From the olivary pretectal nucleus, fibers project to the Edinger-Westphal nucleus of the midbrain. Pre-ganglionic parasympathetic fibers travel with the oculomotor nerve (CN III) to the ciliary ganglion, and post-ganglionic fibers travel to the sphincter muscle of the iris.

Figure Description

Eyes, optic nerves, and thalamus

Key

A	eye
B	optic nerve
C	optic chiasm
D	optic tract
E	midbrain
F	lateral geniculate nucleus
G	superior colliculus

CRANIAL NERVE II

the optic nerve

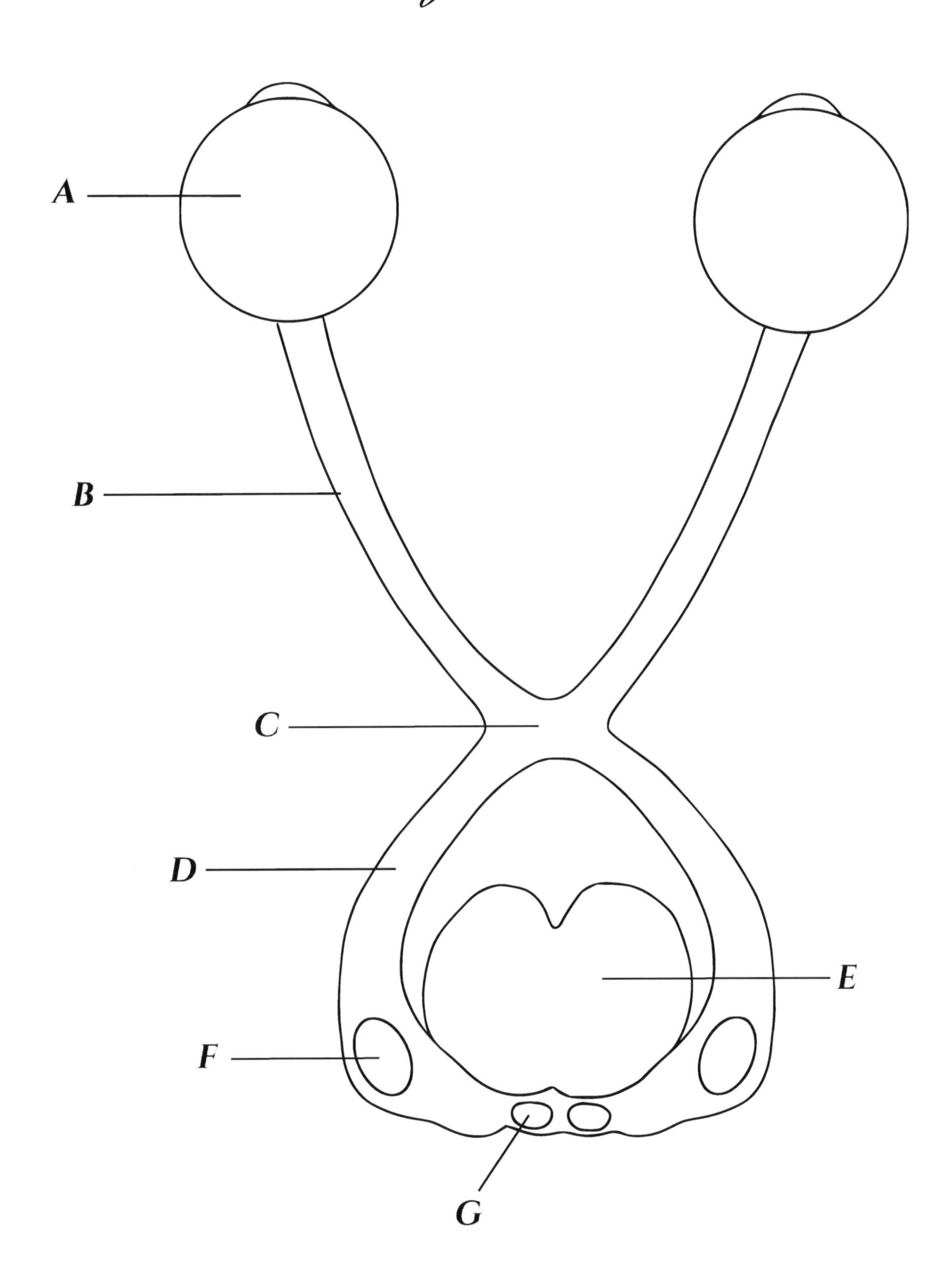

Cranial Nerve III

Cranial nerve III is the oculomotor nerve, which has motor and parasympathetic functions. CN III originates in the mesencephalon, or midbrain. Nuclei include the oculomotor nucleus and Edinger-Westphal nucleus. The nuclei are located in the midbrain at the level of the superior colliculus.

The oculomotor nucleus includes subnuclei for the extraocular muscles it innervates, which include the levator palpebrae superioris, superior rectus, medial rectus, inferior rectus, and inferior oblique muscles. Nerve fibers innervating the medial rectus, inferior rectus, and inferior oblique are from ipsilateral subnuclei. Nerve fibers innervating the superior rectus are from contralateral subnuclei. Nerve fibers innervating the levator palpebrae superioris are bilateral.

Fibers from the subnuclei join together and exit the midbrain, passing through the cavernous sinus. The nerve then divides into superior and inferior divisions, both of which enter the orbit through the superior orbital fissure within the annulus of Zinn. The superior division provides motor innervation to the levator palpebrae superioris and the superior rectus. The inferior division provide motor innervation to the medial rectus, inferior rectus, and inferior oblique. The inferior division also carries parasympathetic fibers.

The cell bodies of pre-ganglionic parasympathetic fibers carried in CN III are located in the Edinger-Westphal nucleus of the midbrain. The pre-ganglionic fibers travel with the inferior division of CN III , branching off within the orbit and synapsing in the ciliary ganglion. The post-ganglionic fibers enter the eye as the short ciliary nerves, synapsing on the ciliary muscle for accommodation, and the sphincter muscle of the iris for miosis (pupil constriction).

Figure Description

Right oculomotor nerve and bony orbit, lateral view with zygomatic bone removed and lateral rectus muscle resected

Key

A levator palpebrae superioris muscle
B superior rectus muscle
C lateral rectus muscle, resected
D inferior rectus muscle
E medial rectus muscle
F superior oblique muscle
G inferior oblique muscle
H optic nerve
I annulus of Zinn
J frontal bone
K oculomotor nerve
L superior division of oculomotor nerve
M inferior division of oculomotor nerve
N ciliary ganglion
O short ciliary nerves

Cranial Nerve III

the oculomotor nerve

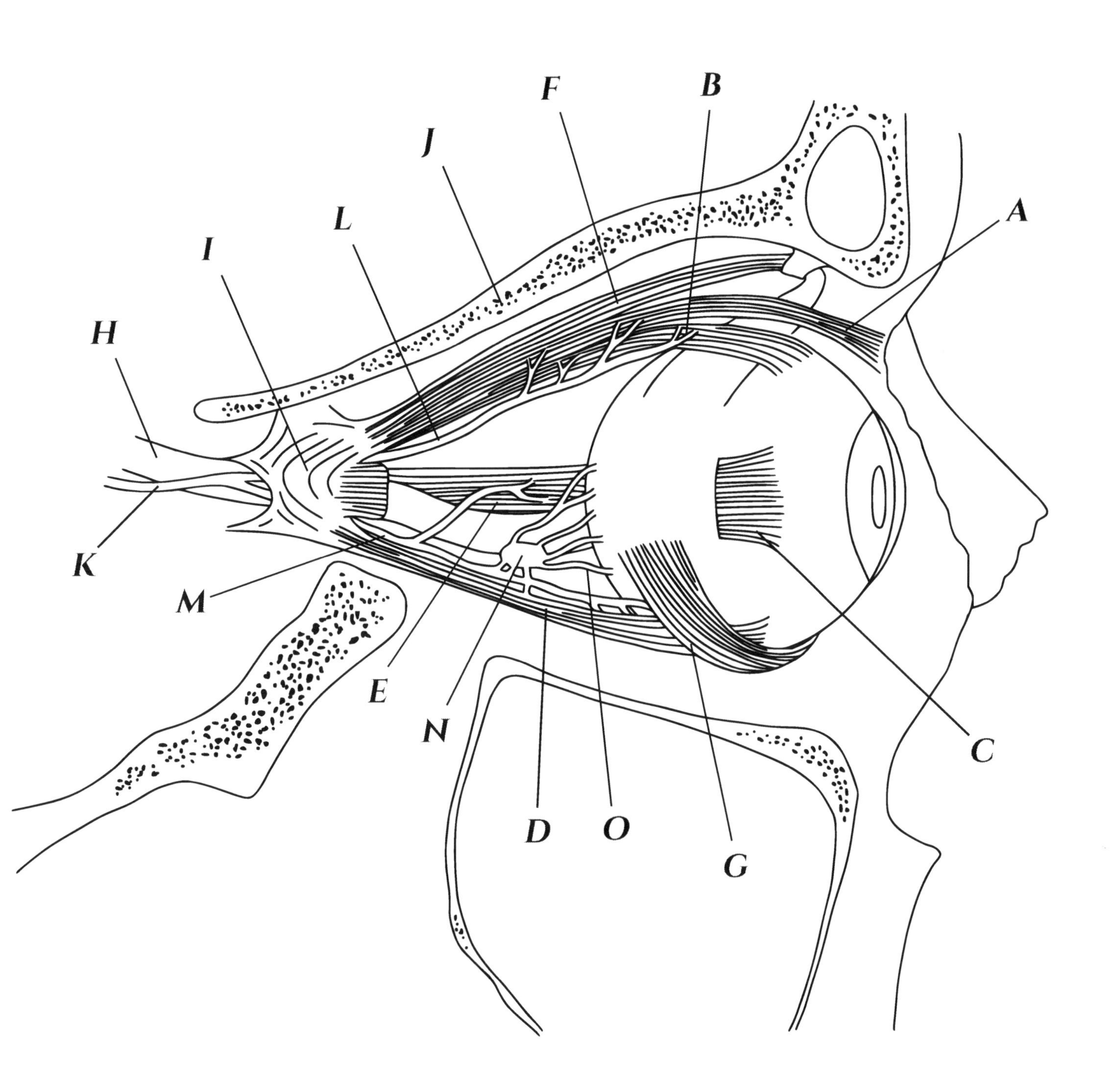

Cranial Nerve IV

Cranial nerve IV is the trochlear nerve, which has only a motor function. CN IV innervates a single extraocular muscle, the superior oblique. It carries general somatic efferent information.

CN IV originates in the mesencephalon, or midbrain, at the trochlear nucleus. The nucleus is located in the midbrain at the level of the inferior colliculus. Fibers of the trochlear nerve decussate, then exit the midbrain dorsally; CN IV is the only cranial nerve to exit the brainstem dorsally. CN IV is the longest and thinnest cranial nerve, being approximately 75 mm and having the fewest number of axons. As it travels anteriorly, it passes through the cavernous sinus along the lateral wall, inferior to CN III. It enters the orbit through the superior orbital fissure, outside of the annulus of Zinn. The nerve then travels superiorly and anteriorly to innervate the belly of the superior oblique muscle.

Figure Description

Right trochlear nerve and bony orbit, lateral view with zygomatic bone removed and lateral rectus muscle resected

Key

A levator palpebrae superioris muscle
B superior rectus muscle
C lateral rectus muscle, resected
D inferior rectus muscle
E medial rectus muscle
F superior oblique muscle
G inferior oblique muscle
H optic nerve
I annulus of Zinn
J frontal bone
K trochlear nerve

Cranial Nerve IV

the trochlear nerve

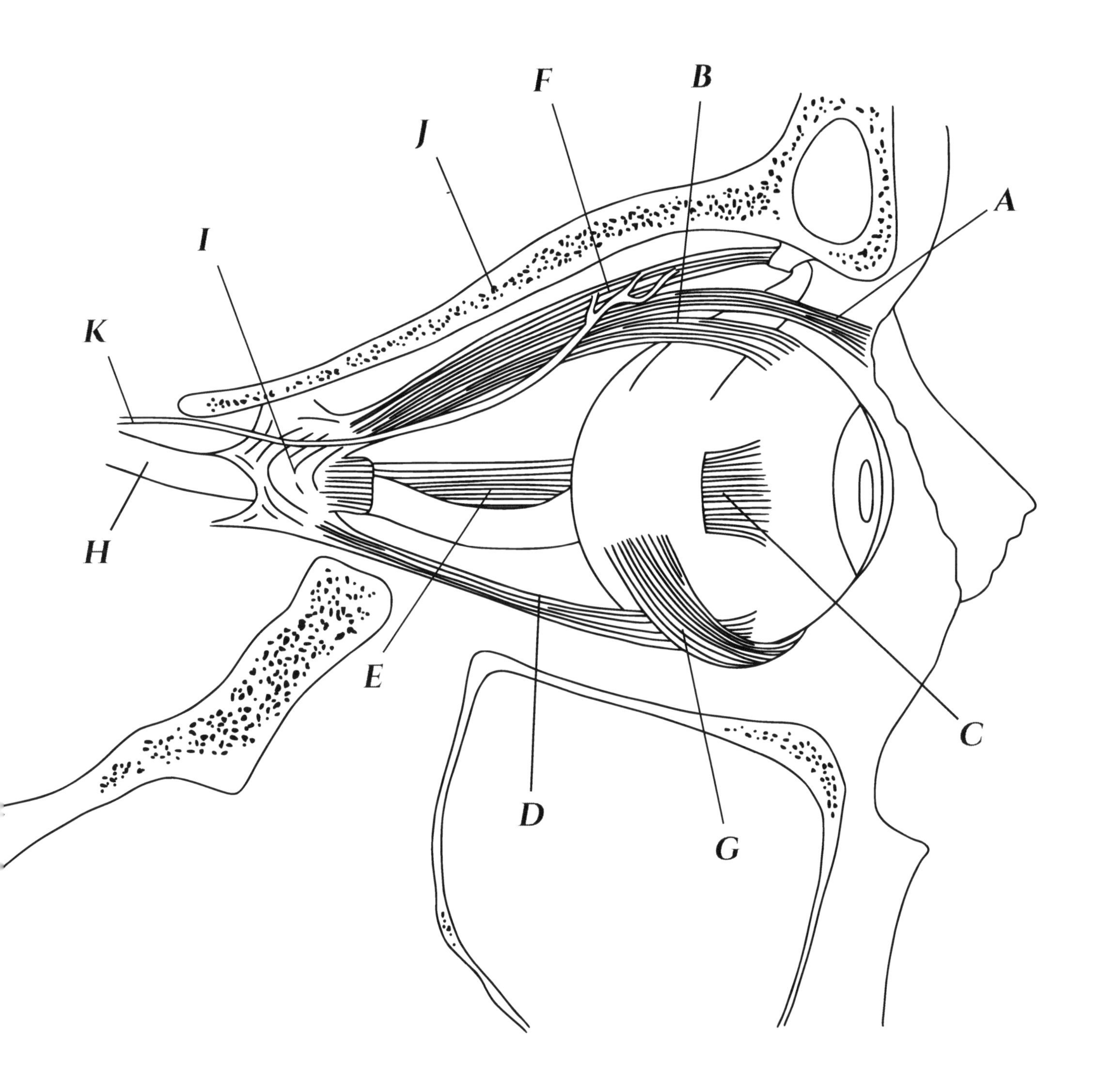

Cranial Nerve V

Cranial nerve V is the trigeminal nerve, which has sensory and motor functions, carrying general somatic afferent and special visceral efferent information. CN V has four nuclei, including the mesencephalic nucleus, motor nucleus, principal pontine sensory nucleus, and spinal nucleus. The fibers converge and emerge from the pons as two roots, a larger sensory root and smaller motor root. The fibers expand within the trigeminal ganglion in the middle cranial fossa, and branch into three divisions, ophthalmic (CN V1), maxillary (CN V2), and mandibular (CN V3). All divisions carry sensory information, and V3 also carries motor information. Here, we will focus on V1 and V2, which have branches that enter the orbit.

CN V1 travels through the cavernous sinus along the lateral wall, inferior to CN IV. It then branches into three nerves, the lacrimal, frontal, and nasociliary nerves. These nerves provide sensory innervation to the superior 1/3 of the face. The lacrimal and frontal nerves pass through the superior orbital fissure outside of the annulus of Zinn, and the nasociliary nerve passes through the superior orbital fissure inside of the annulus of Zinn. The frontal nerve branches into the supratrochlear and supraorbital nerves, providing sensory innervation to the eyelid, conjunctiva, and skin of forehead. The nasociliary nerve branches into the posterior and anterior ethmoidal nerves, inferior trochlear nerve, and long and short ciliary nerves. The long and short ciliary nerves provide sensory innervation to the eye, including the cornea. Sensory fibers of the short ciliary nerve pass through, but do not synapse in, the ciliary ganglion. The lacrimal nerve provides sensory innervation to the lacrimal gland, conjunctiva, and eyelid.

CN V2 provides sensory innervation to the middle 1/3 of the face. CN V2 travels through the cavernous sinus along the lateral wall, inferior to CN V1, then passes through the foramen rotundum. It then gives off numerous sensory branches. Of most relevance to the eye are the infraorbital and zygomatic nerves, which enter the orbit through the inferior orbital fissure outside of the annulus of Zinn. The infraorbital nerve travels on the floor of the orbit in the infraorbital groove, passes through the infraorbital canal, and emerges onto the cheek from the infraorbital foramen. The zygomatic nerve travels along the lateral wall of the orbit and divides into the zygomaticotemporal and zygomaticofacial nerves. These nerves pass through the zygomaticotemporal and zygomaticofacial foramina to supply the lateral aspect of the cheek.

Figure Description

Lateral view of brainstem and right trigeminal nerve and bony orbit, highlighting branches of the ophthalmic and maxillary divisions

Key

A	pons
B	trigeminal nerve
C	trigeminal ganglion
D	ophthalmic division of trigeminal nerve
E	frontal nerve
F	supraorbital nerve
G	supratrochlear nerve
H	lacrimal nerve
I	nasociliary nerve
J	posterior ethmoidal nerve
K	anterior ethmoidal nerve
L	long ciliary nerve
M	ciliary ganglion
N	short ciliary nerve
O	maxillary division of trigeminal nerve
P	zygomatic nerve
Q	infraorbital nerve
R	posterior superior alveolar nerve
S	mandibular division of trigeminal nerve
T	superior cervical ganglion
U	internal carotid artery
V	carotid plexus
W	lacrimal gland

Cranial Nerve V

the trigeminal nerve

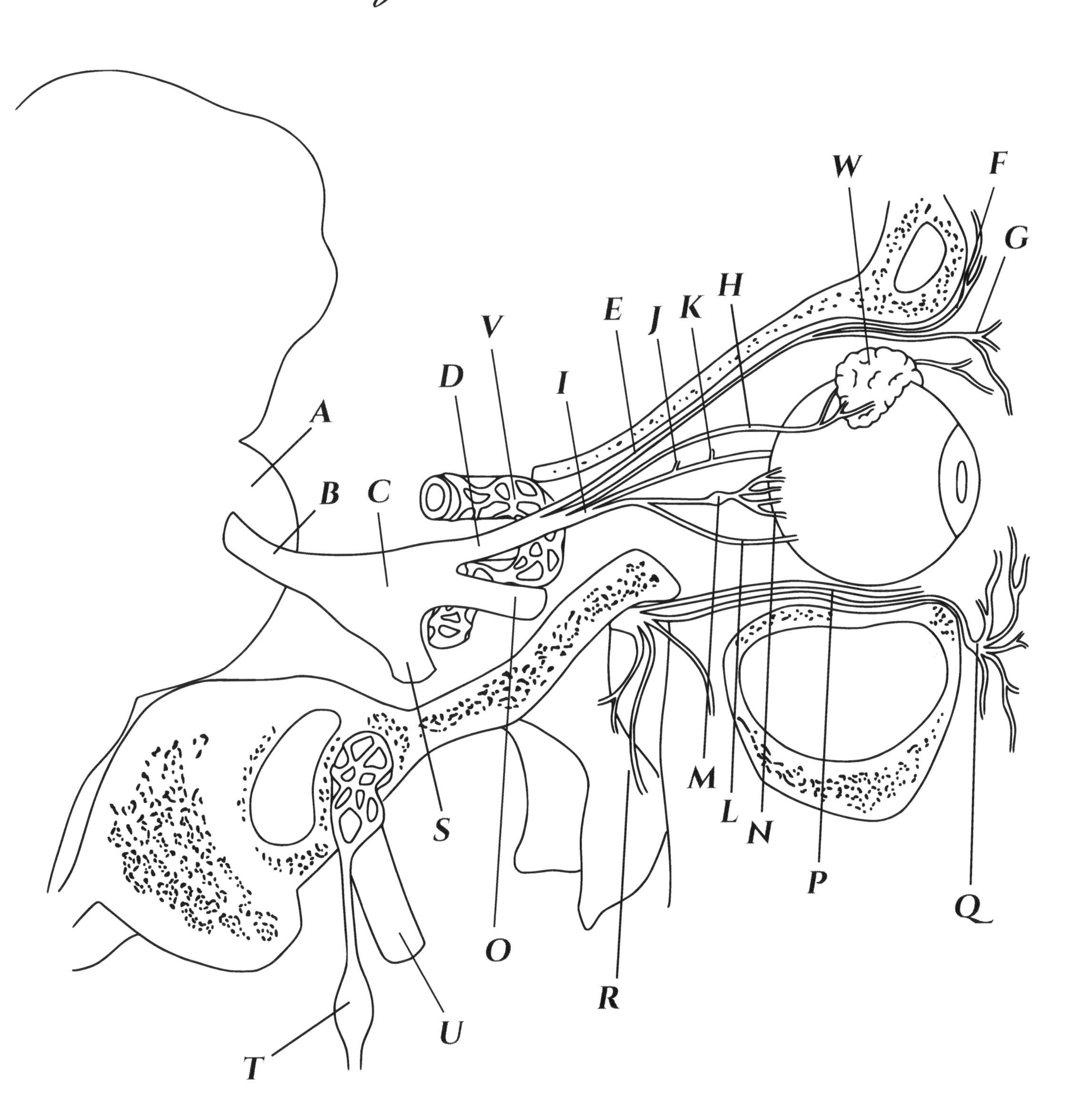

Cranial Nerve VI

Cranial nerve VI is the abducens nerve, which has only a motor function. CN VI innervates a single extraocular muscle, the lateral rectus. It carries general somatic efferent information.

CN VI originates in the metencephalon, or pons, at the abducens nucleus, which is located on the floor of the fourth ventricle at the level of the facial colliculus. The nerve emerges from the junction of the pons and medulla, then courses over the petrous portion of the temporal bone. It passes through the cavernous sinus, then enters the orbit through the superior orbital fissure, inside of the annulus of Zinn. The nerve then travels laterally and anteriorly to innervate the belly of the lateral rectus muscle.

Figure Description

Right abducens nerve and bony orbit, lateral view with zygomatic bone removed and lateral rectus muscle reflected

Key

A	levator palpebrae superioris muscle
B	superior rectus muscle
C	lateral rectus muscle, reflected
D	inferior rectus muscle
E	medial rectus muscle
F	superior oblique muscle
G	inferior oblique muscle
H	optic nerve
I	annulus of Zinn
J	frontal bone
K	abducens nerve

Cranial Nerve VI

the abducens nerve

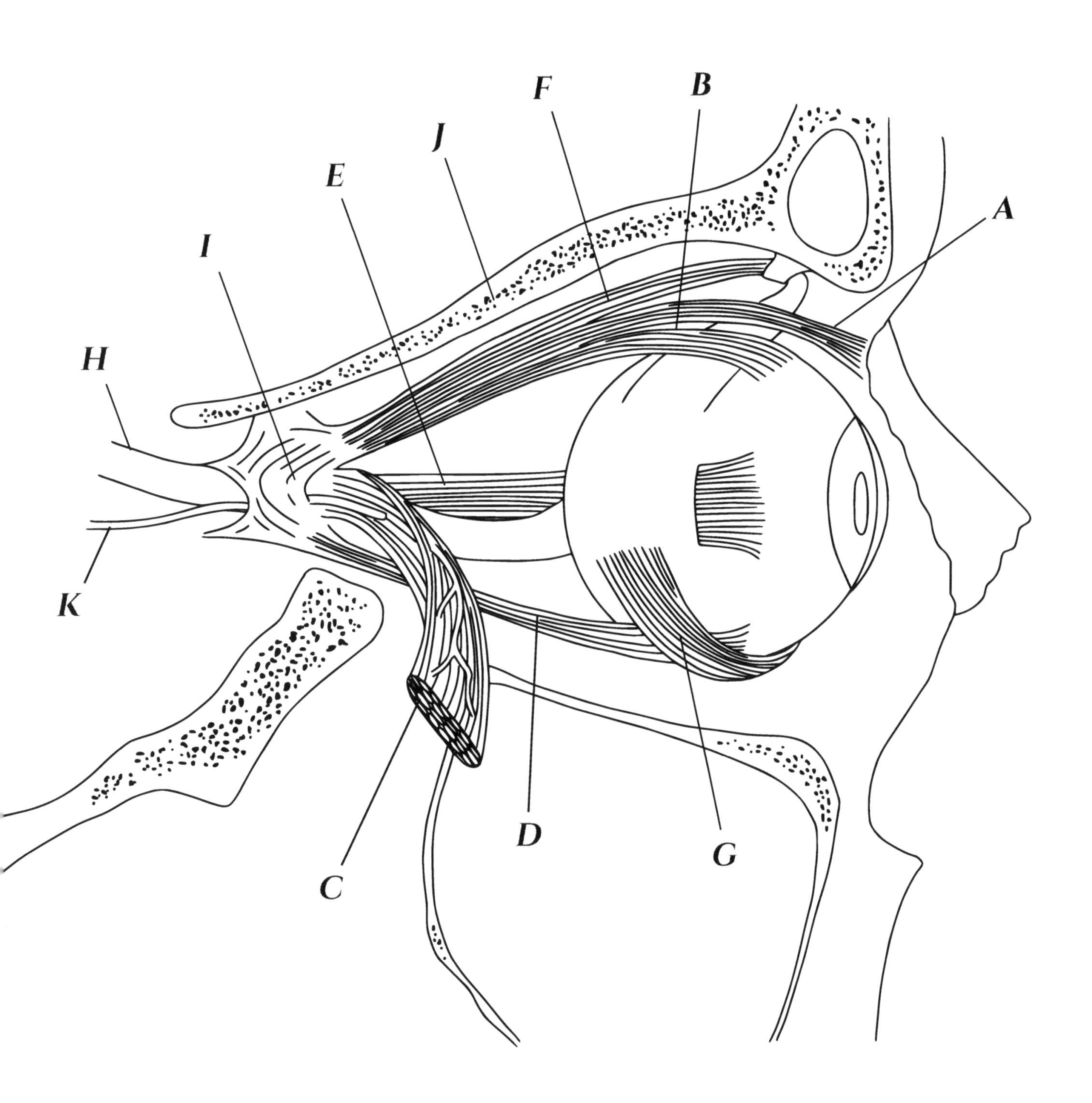

Cranial Nerve VII, Parasympathetic Components

Cranial nerve VII is the facial nerve, which has motor, sensory, and parasympathetic functions, carrying general and special visceral efferent and afferent information. CN VII has four nuclei, including the facial nucleus (motor), superior salivatory nucleus (parasympathetic efferent), spinal nucleus of trigeminal (sensory), and the nucleus of tractus solitarius (sensory). The fibers of the nuclei converge, and the nerve emerges from the junction of the pons and medulla as three roots, a larger motor root and smaller parasympathetic and sensory roots. After emerging from the brainstem, the roots pass through the internal acoustic meatus and into the facial canal, where the roots fuse to form the facial nerve proper.

This page focuses on the parasympathetic component of CN VII, and the following page focuses on the motor and sensory components. The parasympathetic fibers innervate the lacrimal, nasal, palatine, pharyngeal, sublingual, and submandibular glands. Parasympathetic stimulation increases secretion of these glands.

The parasympathetic nucleus is the superior salivatory nucleus. Pre-ganglionic parasympathetic fibers emerge as the nervus intermedius. The fibers pass through, but do not synapse in, the geniculate ganglion, and branch into two pathways. One pathway travels within the greater petrosal nerve, joining the deep petrosal nerve and becoming the vidian nerve (aka nerve of the pterygoid canal). These fibers pass through the pterygoid canal and synapse in the pterygopalatine ganglion. The post-ganglionic parasympathetic fibers join the zygomatic branch of V2, which sends communicating branches to the lacrimal nerve of V1 and innervates the lacrimal gland. Other post-ganglionic parasympathetic fibers travel as the major superficial petrosal nerve to innervate the nasal glands.

The second pathway travels inferiorly as the mixed facial nerve proper, passing through the facial canal. The nerve gives off the chorda tympani, which travels through the middle ear and exits via the petrotympanic fissure, entering the infratemporal fossa. The chorda tympani travels with the lingual nerve of CN V3. The pre-ganglionic parasympathetic fibers synapse in the submandibular ganglion, and post-ganglionic fibers innervate the submandibular and sublingual glands.

Figure Description

Lateral view of brainstem, facial nerve, and bony orbit, highlighting the parasympathetic components

Key

A pons
B superior salivatory nucleus
C nervus intermedius
D geniculate ganglion
E greater petrosal nerve
G nerve of pterygoid canal, aka vidian nerve
H pterygopalatine ganglion
I rami ocularis
J lacrimal gland
K major superficial petrosal nerve
L nasal mucosa glands
M chorda tympani nerve
N submandibular ganglion
O submandibular gland
P sublingual gland
Q frontal bone
R superior cervical ganglion
S internal carotid artery

CRANIAL NERVE VII

the facial nerve, parasympathetic component

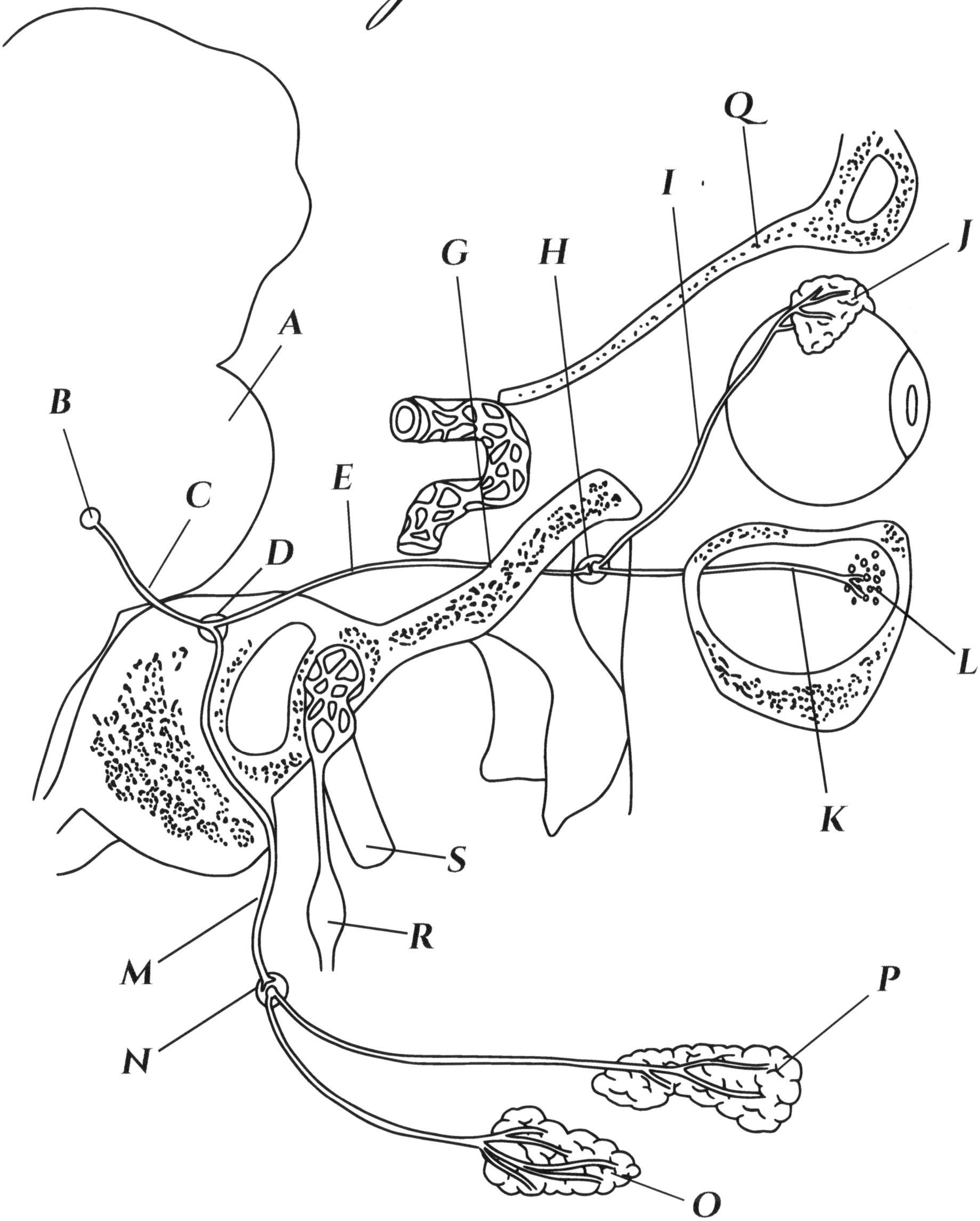

Cranial Nerve VII, Motor & Sensory Components

Cranial nerve VII, the facial nerve, is introduced in the previous page. The current page focuses on the motor and sensory components of CN VII, and the previous page focuses on the parasympathetic component. The motor functions include innervation of the muscles of facial expression, as well as the stapedius, posterior auricular, digastric (posterior belly), and stylohyoid muscles. The sensory function includes taste of the anterior 2/3 of the tongue.

The motor nucleus of CN VII is the facial nucleus. Motor fibers emerge from the junction of the pons and medulla and pass through the internal acoustic meatus to the facial canal of the temporal bone. Here, the first motor branch is given off - the stapedius nerve to the stapedius muscle of the middle ear.

The motor root then exits the skull through the stylomastoid foramen. The next motor branch is the posterior auricular nerve, which provides innervation to the muscles around the ear. Then, the nerves to the posterior belly of the digastric muscle and the stylohyoid muscle branch off.

The motor root then passes through (but does not innervate) the parotid gland, where the five terminal motor nerves branch to supply the muscles of the face and neck. From superior to inferior, these branches are the 1) temporal branch, supplying the occipitofrontalis (frontal belly), procerus, orbicularis oculi and corrugator supercilii muscles, 2) zygomatic branch, supplying the orbicularis oculi muscle, 3) buccal branch, supplying the orbicularis oris, buccinator and zygomaticus muscles, 4) mandibular branch, supplying the mentalis muscle, and 5) cervical branch, supplying the platysma.

The special sensory function of CN VII is taste. The sensory nucleus is the nucleus solitarius. Sensory fibers are carried in the chorda tympani, which branches from the motor root in the geniculate ganglion. The choroid tympani carries both sensory taste fibers from the tongue and parasympathetic fibers to the sublingual and submandibular glands. The chorda tympani travels through the middle ear and exits via the petrotympanic fissure, entering the infratemporal fossa. The chorda tympani travels with the lingual nerve of CN V3, then the taste fibers branch off to innervate the anterior 2/3 of the tongue.

Figure Description

Terminal motor branches of the facial nerve, lateral view of face

Key

A	temporal branch
B	zygomatic branch
C	buccal branch
D	mandibular branch
E	cervical branch
F	posterior auricular nerve

The general somatic afferent function of CN VII is sensory innervation to the skin of the posterior ear, specifically the external auditory meatus, the tympanic membrane, and skin over the mastoid process and pinna of the ear. These fibers originate in the spinal trigeminal nucleus.

CRANIAL NERVE VII

the facial nerve, motor component

Cranial Nerve VIII & the Vestibulo Ocular Reflex

Cranial nerve VIII, the vestibulocochlear nerve, does not directly innervate the eye or orbit, but is part of the vestibulo ocular reflex. The vestibulo ocular reflex serves to stabilize images on the retina during head movement. CN VIII functions in equilibrium and balance (the vestibular component) and in hearing (the cochlear, or auditory component). Here, we will focus on the vestibular component that modulates the vestibulo ocular reflex.

Receptors of the vestibular nerve are in the ampullae of the semicircular canals of the inner ear. Cristae located in the ampullae contain hair cells that respond to rotational acceleration of the head. Hair cells in the saccule and utricle respond to linear acceleration.

Information is carried along the vestibular nerve, with the cell bodies being in the vestibular ganglion. The fibers join CN VII and pass through the internal acoustic meatus. CN VIII enters the brainstem at the junction of the pons, medulla, and cerebellum, known as the cerebellopontine angle. There are four vestibular nuclei in the pons. From these nuclei, fibers decussate to the contralateral abducens nucleus, where the signal splits into two pathways. One pathway follows CN VI (the abducens nerve) to the lateral rectus muscle. The other pathway travels from the abducens nucleus to the oculomotor nucleus on the contralateral side (i.e. the side where the signal originated). From here, the signal travels in CN III (the oculomotor nerve) to the medial rectus. There are also torsional components of the vestibulo ocular reflex.

Figure Description

Cranial nerves VIII, III, and VI and their effector organs related to the vestibulo ocular reflex

Key

A lateral semicircular canal
B posterior semicircular canal
C anterior semicircular canal
D cochlea
E utricle
F saccule
G vestibular ganglion
H vestibular root, cranial nerve VIII
I auditory root, cranial nerve VIII
J vestibular nucleus
K abducens nucleus
L oculomotor nucleus
M oculomotor nerve
N abducens nerve
O medial rectus muscle
P lateral rectus muscle

Cranial Nerve VIII

the vestibular nerve

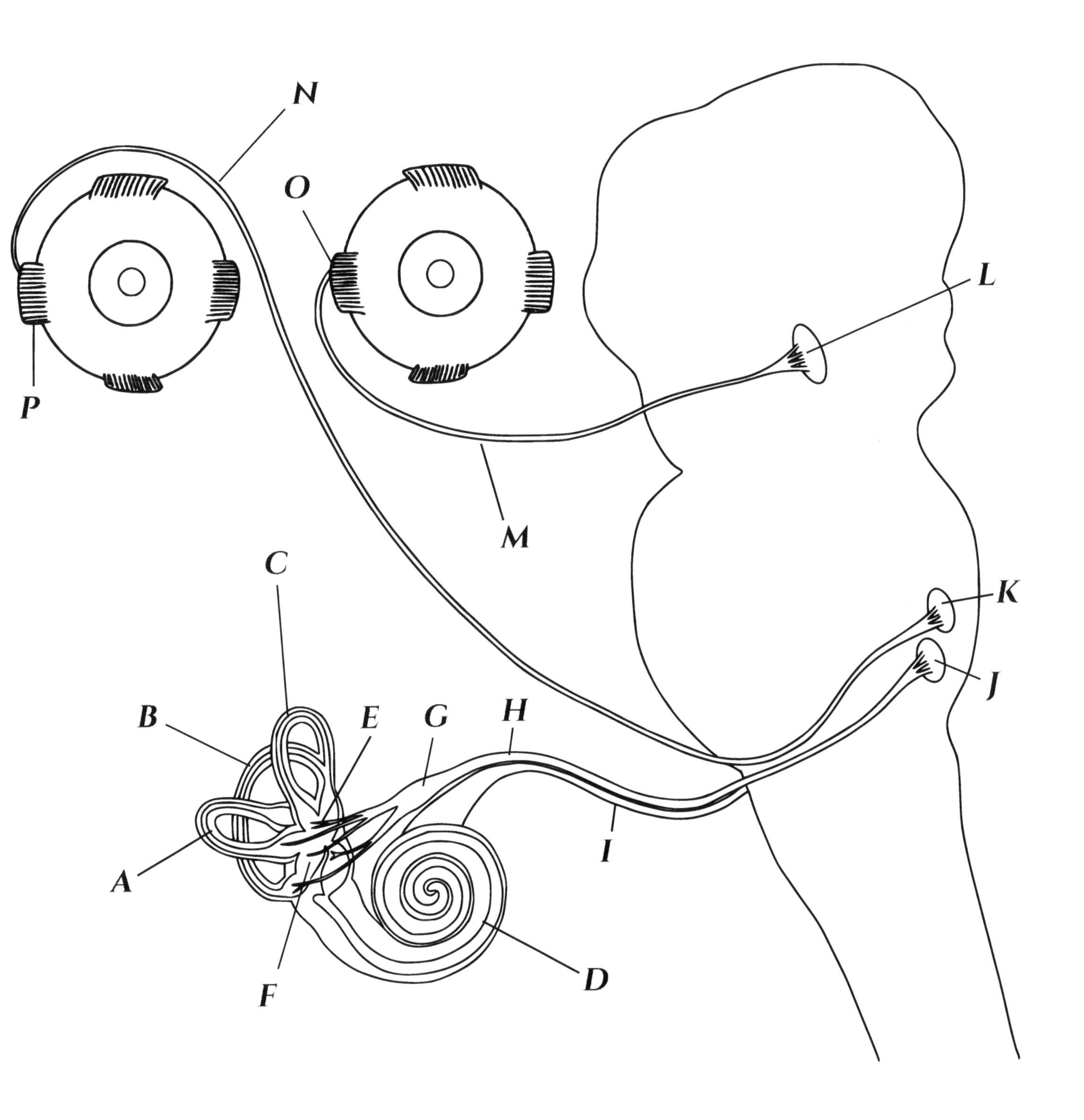

Ciliary Nerves

Innervation to the eye is carried by the ciliary nerves, which are in two groups of nerves, long and short, piercing the globe near the optic nerve and traveling within the suprachoroidia to numerous structures of the eye.

The long ciliary nerves carry sensory and sympathetic information. There are generally two long ciliary nerves branching off of the nasociliary nerve as it crosses over the optic nerve. The nasociliary nerve is a branch of the ophthalmic nerve (CN V1). The long ciliary nerves pierce the sclera on either side of the optic nerve and travel along the horizontal meridian to the anterior segment, giving off numerous branches. Sensory information is carried from the cornea, iris, ciliary body, and sclera in the long ciliary nerves to the nasociliary nerve. Sympathetic information reaches the long ciliary nerves via the superior cervical ganglion. Post-ganglionic fibers travel from the superior cervical ganglion in the carotid plexus to CN V1, then branch into the nasociliary nerve then into the long ciliary nerves. Some sympathetic fibers branch from the nasociliary nerve to run through (but not synapse in) the ciliary ganglion and join the short ciliary nerves. Sympathetic input in the eye leads to mydriasis, or pupil dilation, via the iris dilator muscle.

The short ciliary nerves carry primarily parasympathetic information, with a minor sympathetic component. There are 6-8 short ciliary nerves that pierce the sclera around the optic nerve. Parasympathetic information carried in the short ciliary nerves originates in the Edinger-Westphal nucleus of the midbrain. The pre-ganglionic fibers travel with CN III (the oculomotor nerve) and synapse in the ciliary ganglion. The post-ganglionic fibers of the ciliary ganglion are the short ciliary nerves, which pierce the sclera, travel to the anterior segment within the suprachoroidia, and innervate the ciliary muscle and iris sphincter muscle.

Figure Description

Superior view of right eye, outer tunic partially resected, highlighting the ciliary nerves, adapted from Wood, 1913

Key

A medial long ciliary nerve
B lateral long ciliary nerve
C short ciliary nerves
D parasympathetic nerves to ciliary muscle
E sympathetic nerves to iris dilator muscle
F parasympathetic nerves to iris sphincter muscle
G sensory nerves of cornea

Ciliary Nerves

long and short

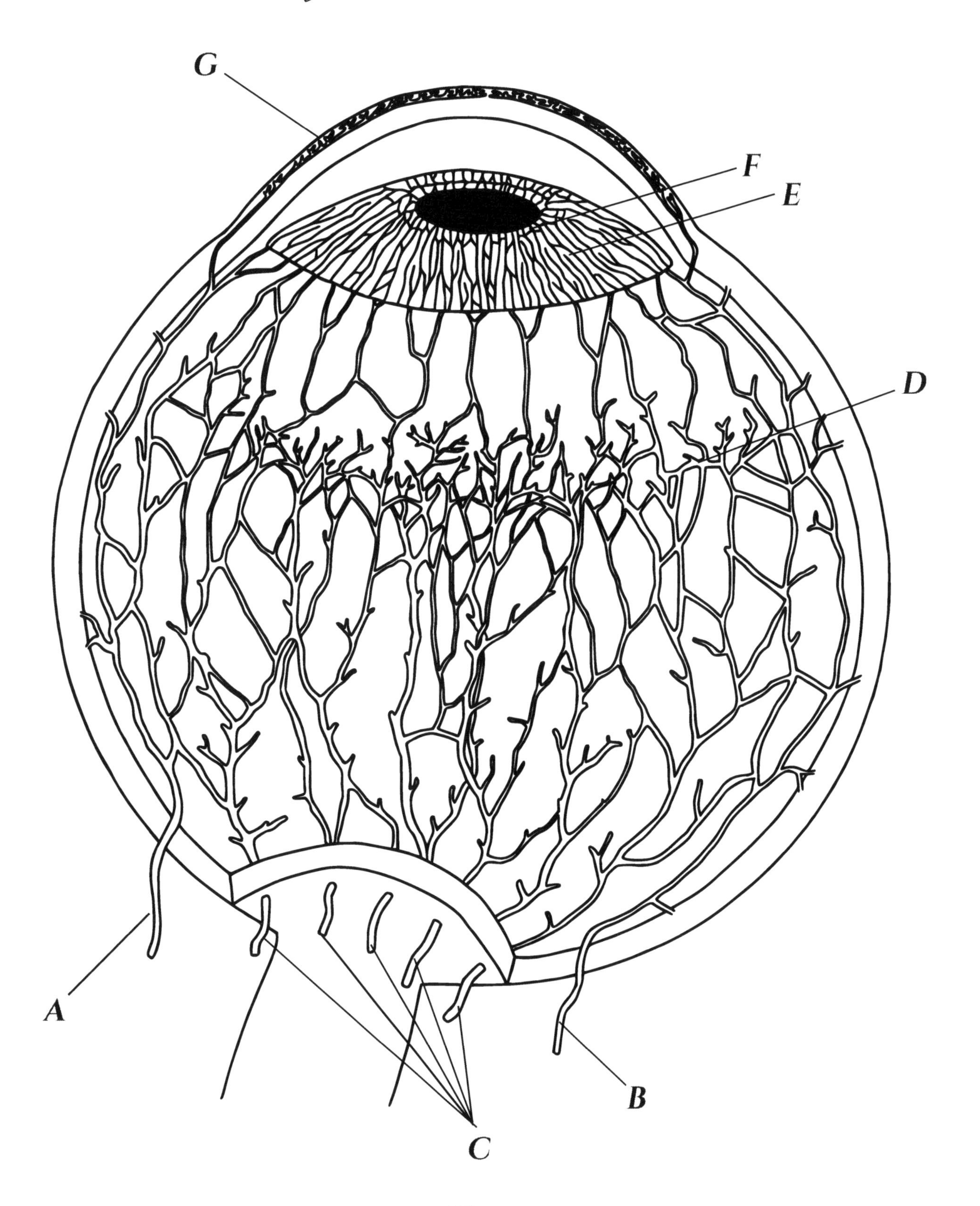

Autonomic Pathways

Autonomic input to the eye includes sympathetic innervation to the dilator muscle, and parasympathetic innervation to the ciliary muscle and sphincter muscle. Additionally, the tarsal (Mueller) muscle of the eyelid receives sympathetic input and the lacrimal gland receives parasympathetic input. Ocular blood flow, including vasculature of the optic nerve, choroid, retina, iris, and ciliary body, also has autonomic input.

Sympathetic input to the eye leads to mydriasis, or pupil dilation. Pre-ganglion sympathetic neurons are located in the intermediolateral cell column in the spinal cord, in the 8th cervical (C8) to 2nd thoracic (T2) spinal segments. The axons emerge from the spinal cord and enter the sympathetic chain, traveling superiorly to synapse in the superior cervical ganglion. Post-ganglionic sympathetic axons travel from the superior cervical ganglion in the carotid plexus, a network around the carotid artery. The fibers join the nasociliary nervem a branch of the ophthalmic nerve (CN V1), and enter the eye via two pathways: 1) with the long ciliary nerves and 2) passing through (but not synapsing in) the ciliary ganglion and joining the short ciliary nerves. Pre-ganglionic neurons are cholinergic and release acetylcholine, and post-ganglionic neurons are adrenergic and release norepinephrine.

Parasympathetic input to the eye induces accommodation via the ciliary muscle and miosis (pupil constriction) via the iris sphincter muscle. Parasympathetic input originates at the Edinger-Westphal nucleus of the midbrain. Pre-ganglionic fibers emerge from the brainstem with CN III (the oculomotor nerve), travel through the cavernous sinus, and enter the orbit with the inferior division of CN III through the superior orbital fissure, within the annulus of Zinn. The parasympathetic fibers branch off and synapse in the ciliary ganglion. Post-ganglionic fibers enter the eye with the short ciliary nerves and travel to the anterior segment to innervate the sphincter and ciliary muscles. Both pre-ganglionic and post-ganglionic fibers are cholinergic, releasing acetylcholine at their terminals.

Figure Description

Left bony orbit, lateral view with zygomatic bone removed, adapted from Gray, 1918

Key

A thalamus
B midbrain
C Edinger-Westphal nucleus
D oculomotor nerve
E pre-ganglionic parasympathetic fibers
F ciliary ganglion
G short ciliary nerves, post-ganglionic parasympathetic fibers
H descending fibers
I thoracic spinal cord
J pre-ganglionic sympathetic fibers
K sympathetic chain
L inferior cervical ganglion
M middle cervical ganglion
N superior cervical ganglion
O carotid plexus
P trigeminal ganglion
Q ophthalmic division of trigeminal nerve
R long ciliary nerves, post-ganglionic sympathetic fibers

AUTONOMIC PATHWAYS

sympathetic and parasympathetic

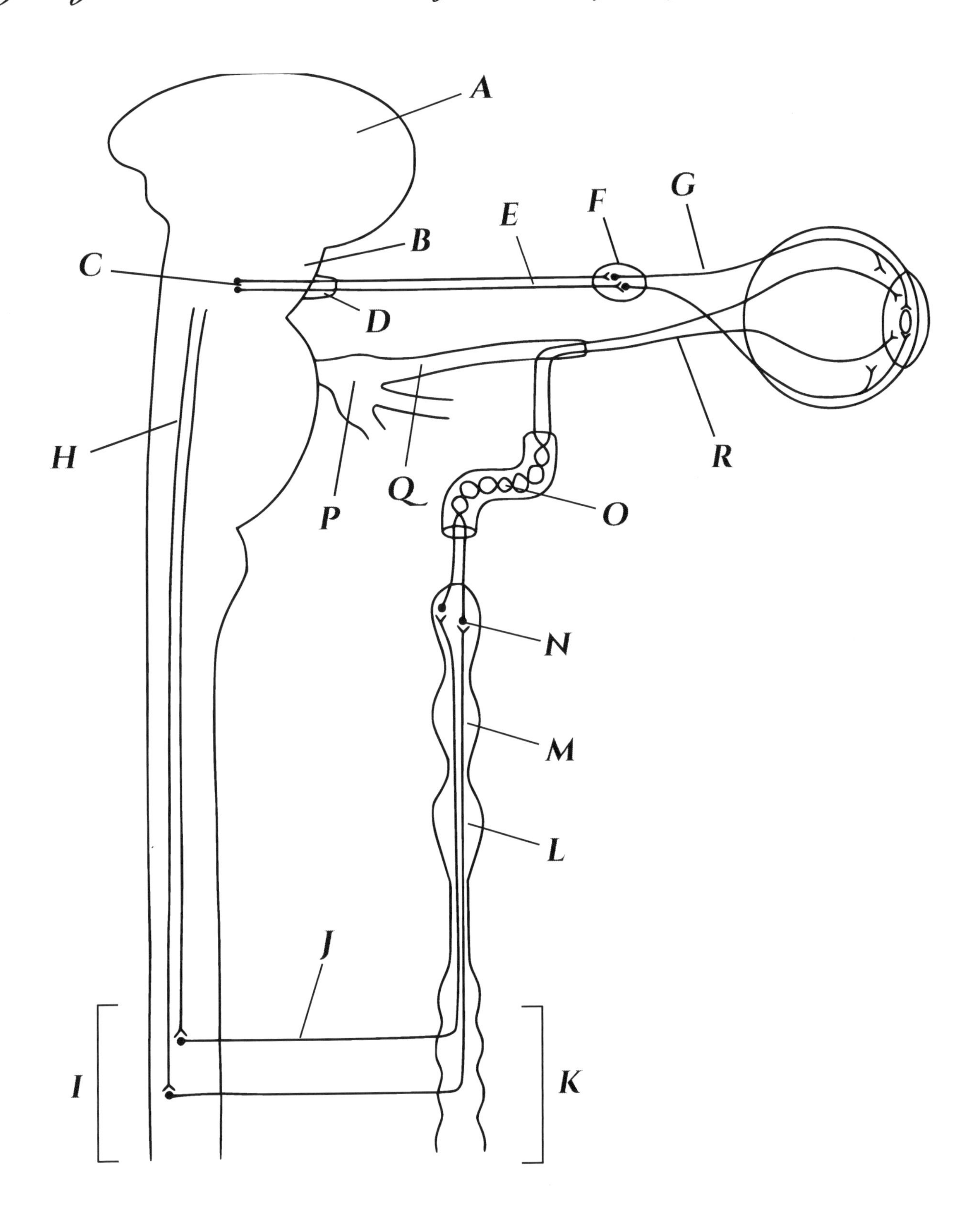

Image Forming Pathway

The image-forming pathway of the visual system refers to the detection of light and subsequent perception of seeing. The pathway begins at the photoreceptors, where light energy is converted to electrical potentials that pass to bipolar cells, then retinal ganglion cells (RGCs). Axons of RGCs converge at the optic nerve head and exit the eye as the optic nerve. After entering the middle cranial fossa, just over half of the fibers, those from the nasal retina, decussate (cross over) at the optic chiasm. Fibers posterior to the optic chiasm make up the optic tract, which synapses in the lateral geniculate nucleus (LGN) of the thalamus. From the LGN, fibers form the optic radiations, which convey information to the visual cortex of the occipital lobe.

The optic radiations are also called the geniculo-calcarine tract. Visual information is carried in two divisions, upper and lower. The upper division projects to the cuneus and contains input from the superior retinal quadrants (i.e. inferior visual field). The lower division travels anteriorly, forming Meyer's loops, then posteriorly to the lingual gyrus. The lower division contains input from the inferior retina (i.e. superior visual field).

Information along the visual pathway maintains a retinotopic organization, meaning, signals are processed in the cortex in specific locations that can be mapped to a precise location on the retina.

Figure Description

Ventral view of brain, highlighting the image forming pathway from retina to visual cortex

Key

A retina
B optic nerve
C optic chiasm
D optic tract
E lateral geniculate nucleus
F optic radiations
G primary visual cortex

IMAGE FORMING PATHWAY

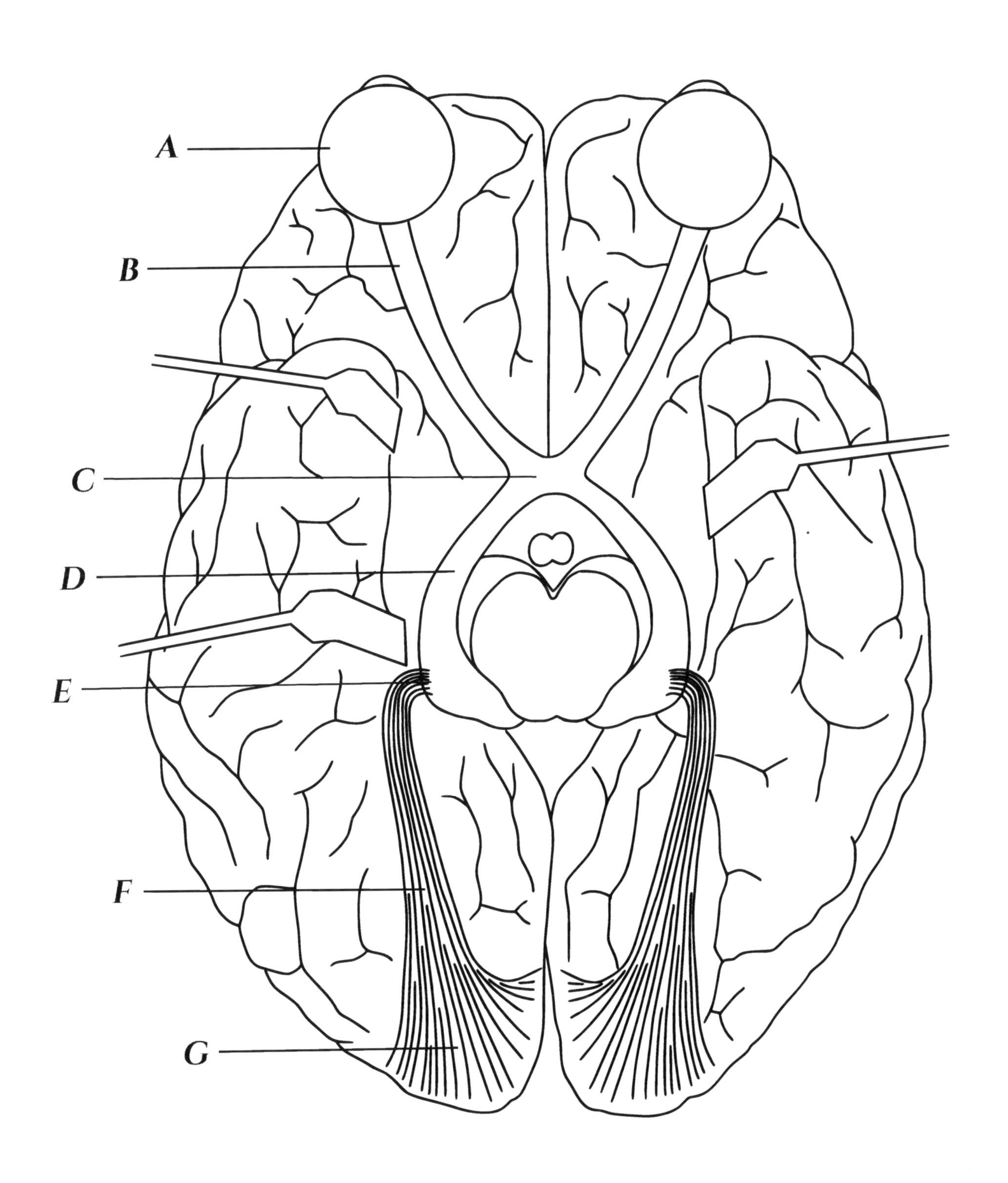

Lateral Geniculate Nucleus

The retinal ganglion cells (RGCs) involved in image perception synapse in the lateral geniculate nucleus (LGN) in the thalamus, a relay center of the visual pathway. The LGN is a paired posterolateral extension of the thalamus. In addition to input from RGCs, the LGN also receives input from the visual cortex, superior colliculus, pretectum, thalamic reticular nuclei, and local LGN interneurons.

The human LGN consists of 6 layers of neurons (i.e. gray matter) with optic fibers (i.e. white matter) in between. The inner two layers (1 and 2) are in the magnocellular pathway, and the outer four layers (3, 4, 5, and 6) are in the parvocellular pathway. These layers are interleaved with cells in the koniocellular pathway.

The LGN conveys information from the contralateral visual field. For example, the left LGN receives input from the temporal left and nasal right retinae, such that it receives input from the right visual field. LGN layers 1, 4, and 6 receive input from crossed fibers (from the contralateral eye), and layers 2, 3, and 5 receive input from uncrossed fibers (from the ipsilateral eye). Additionally, fibers maintain a retinotopic pattern, with fibers from the superior retina synapsing in the medial LGN, and from the inferior retina synapsing in the lateral LGN. Fibers from the macular region synapse on the dorsal aspect of the LGN.

The magnocellular layers of the LGN consist of synapses from magnocellular (M-type or parasol) ganglion cells. The M cells of the LGN have a large cell body, large dendritic tree, and large receptive field. The input originates from rods, as well as cones, functioning in the perception of movement, depth, and small increments in brightness under low illumination. The response properties are rapid and transient.

The parvocellular layers of the LGN consist of synapses from parvocellular (P-type or midget) ganglion cells. The P cells of the LGN have a small cell body, small dendritic tree, and small receptive field. The input originates from red and green cones, functioning in the perception of fine resolution and color. The response properties are slow and sustained.

The koniocellular interlaminar layers of the LGN consist of synapses from small bistratified RGCs. Input originates from blue cones, functioning in color vision.

Figure Description

Cross section of the lateral geniculate nucleus

Key

A dorsal aspect
B ventral aspect
C layer 1, magnocelluar, contralateral
D layer 2, magnocellular, ipsilateral
E layer 3, parvocellular, ipsilateral
F layer 4, parvocellular, contralateral
G layer 5, parvocellular, ipsilateral
H layer 6, parvocellular, contralateral
I koniocellular layers
J hilum

Lateral Geniculate Nucleus

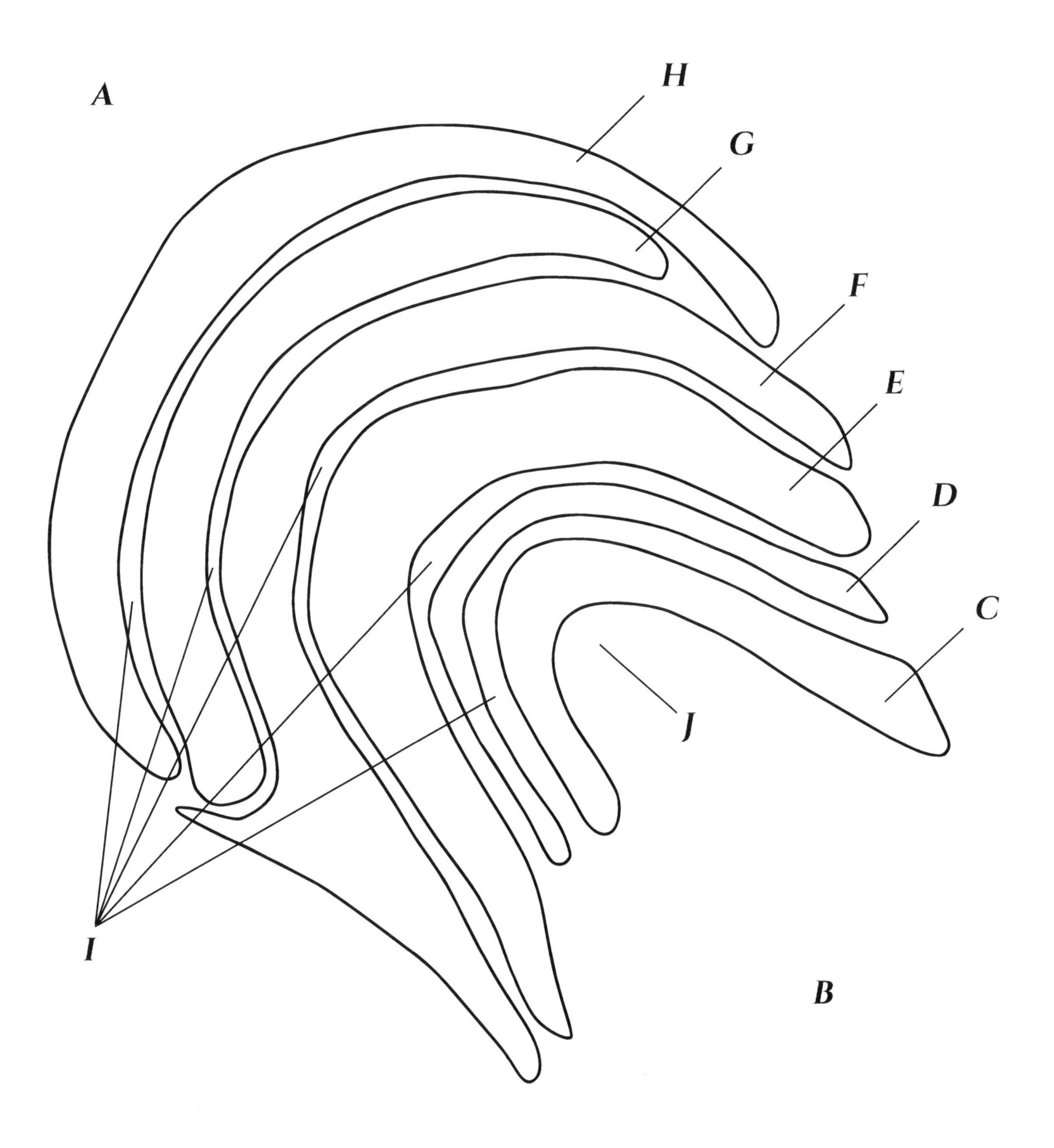

Non-Image Forming Pathways

Non-image forming pathways convey light information from the retina to various brain centers that are not involved in processing images, but rather, regulating pupil size and circadian rhythm entrainment. This information is carried by a subset of retinal ganglion cells that are directly photosensitive via the photopigment melanopsin, known as the intrinsically photosensitive retinal ganglion cells (ipRGCs).

ipRGCs make up 0.5-1.4% of the RGC population in the human retina. ipRGCs express melanopsin and are directly activated by short wavelength light. Additionally, ipRGCs receive input from traditional rod and cone pathways. Some axons of ipRGCs project to the lateral geniculate nucleus, similar to traditional RGCs. However, most axons project to either the olivary pretectal nucleus for pupil size regulation or the suprachiasmatic nucleus for circadian rhythm entrainment.

The afferent pathway for pupil size regulation begins at the retina. Light information travels from the ipRGCs, as well as some traditional RGCs, along the optic nerve to synapse in the olivary pretectal nucleus of the midbrain. Efferent fibers are sent bilaterally to the Edinger-Westphal nucleus of the midbrain, from which pre-ganglionic parasympathetic fibers travel with CN III (the oculomotor nerve) to synapse in the ciliary ganglion, and post-ganglionic fibers travel in the short ciliary nerves to reach the iris sphincter muscle.

The afferent pathway for circadian rhythm entrainment also begins in the retina and is conveyed by the ipRGCs. A subset of ipRGCs synapses in the suprachiasmatic nucleus of the anterior hypothalamus, aka the master clock of the body. The suprachiasmatic nucleus influences numerous processes throughout the body. In particular, a major target is the pineal gland, which secretes melatonin in a diurnal manner that is dependent on light exposure. The suprachiasmatic nucleus has inhibitory projections to the paraventricular nucleus of the hypothalamus. From there, the signal travels to the intermediolateral cell column in the spinal cord and synapses in the superior cervical ganglion. Post-ganglionic fibers synapse in the pineal gland. Given this inhibitory signaling, when light is detected, the pineal gland is inhibited and melatonin is not released. In darkness, inhibition discontinues, and the pineal gland secretes melatonin, signaling the body to prepare for sleep.

Figure Description

Sagittal section of brain with eye in place

Key

A retina
B optic nerve
C suprachiasmatic nucleus
D olivary pretectal nucleus
E pineal gland
F ciliary ganglion
G pituitary gland
H Edinger-Westphal nucleus
I superior cervical ganglion

NON-IMAGE FORMING PATHWAYS

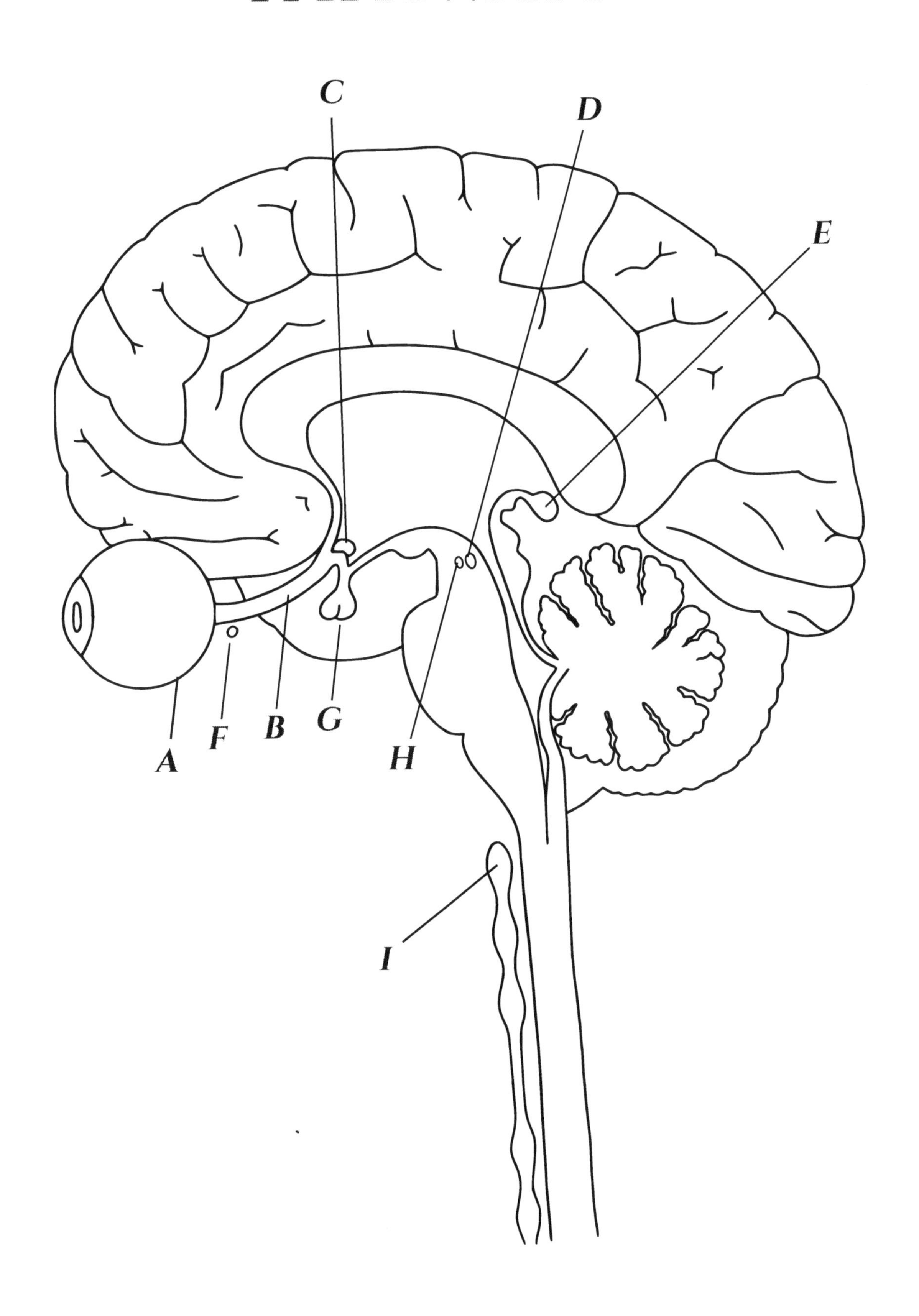

EMBRYOLOGY

Fertilization & Cleavage

Early stages of gestation include fertilization and cleavage. After an oocyte (egg) is released from the ovary, it travels along the fallopian tube, and may be met by sperm in the ampulla region. Hundreds of sperm cells of the 100 million that enter the vaginal canal may reach the oocyte. Generally, a single sperm cell binds to the corona radiata and penetrates the zona pellucida surrounding the oocyte. The acrosome of the sperm fuses with the plasma membrane of the oocyte, the sperm head disconnects from its flagellum, and the sperm nucleus enters the oocyte. Following fusion, the oocyte divides and releases polar bodies. Polar bodies serve to eliminate one half of the diploid chromosome set in the oocyte, leaving a haploid cell. The sperm nucleus fuses with the oocyte nucleus forming a diploid zygote, and a series of cleavage events take place.

Cleavage refers to the repeated mitotic divisions of the zygote, transitioning from a single cell to 2 cells, 4 cells, 8 cells, etc. Each daughter cell is a blastomere. Once the zygote reaches 16 to 32 cells, about 3 days after fertilization, the solid cell mass is called a morula, meaning mulberry.

By day 4, the fluid accumulates in the center of the morula, forming a cavity. The morula transitions into the next stage, the hollow blastula. The inner cavity of the blastocyst is the blastocoel. Cells aggregate at one pole; these cells are the inner cell mass, which ultimately gives rise to the germ layers that develop into an embryo. The outermost layer of cells is the trophoblast.

As the zygote undergoes cleavage, it travels from the fallopian tube to the uterus. At this stage, the zona pellucida begins to degenerate as the embryo prepares for implantation in the uterine wall.

Figure Description

Early embryonic stages, including fertilization, cleavage, morula, and blastula

Key

A fertilization
B 2 cell stage
C 4 cell stage
D 8 cell stage
E morula
F blastula
G ovum
H polar bodies
I corona radiata
J zona pellucida
K nucleus of ovum
L sperm
M nucleus of sperm
N polar bodies
O zona pellucida
P blastomere
Q zona pellucida
R blastomere
S zona pellucida
T blastomere
U zona pellucida
V zona pellucida
W blastoderm/trophoblast
X blastocoel
Y inner cell mass

Fertilization & Cleavage

week 1

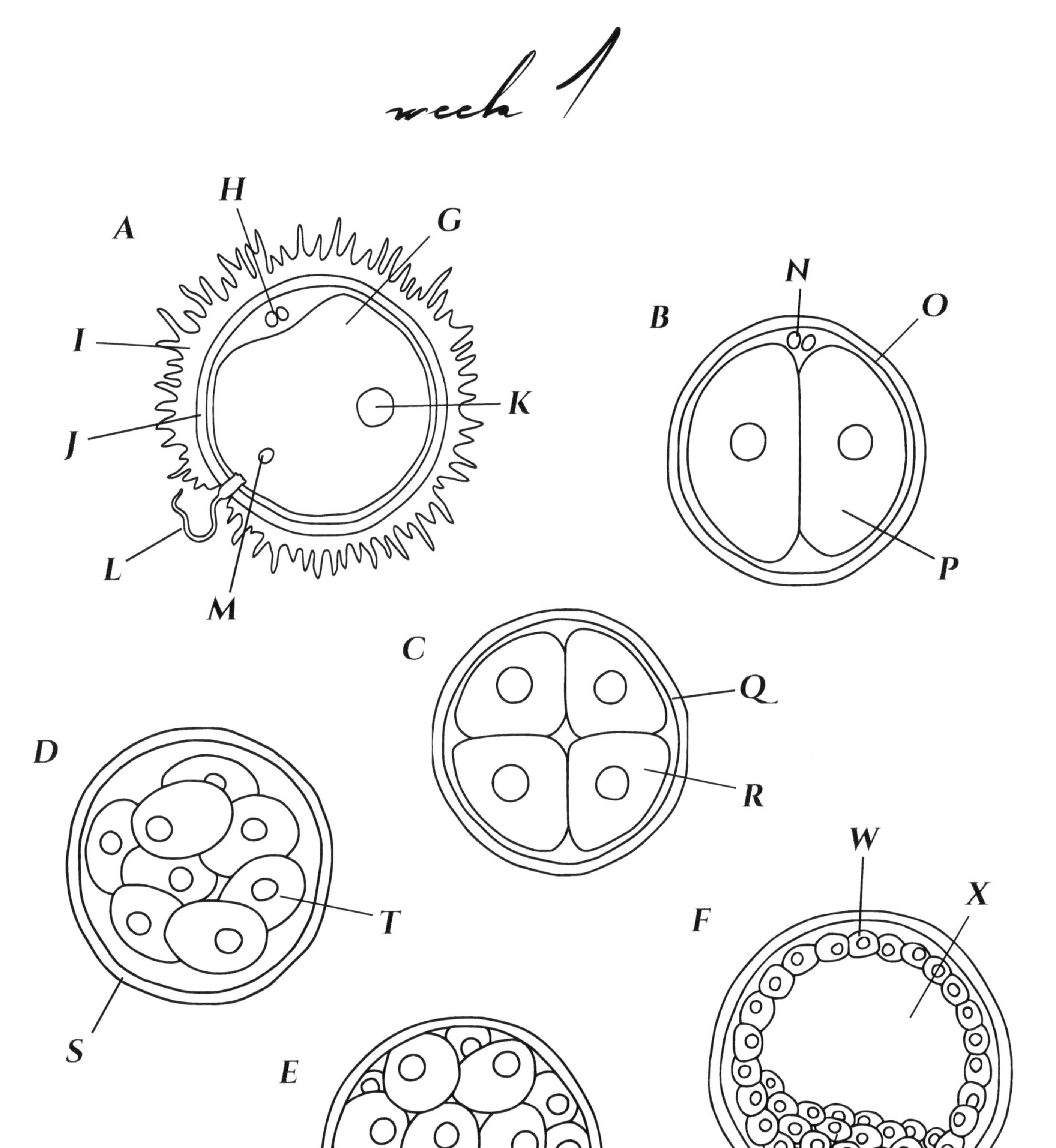

Implantation & Gastrulation

In week two, the embryo reaches the uterine cavity and begins to implant into the uterine wall. First, the trophoblast layer surrounding the blastocyst merges with the uterine lining. The trophoblast has two layers, the inner cytotrophoblast and outer syncytiotrophoblast. The syncytiotrophoblast invades the uterine lining, making direct contact with the maternal blood. The syncytiotrophoblast begins to form lacunae that merge with the mother's blood vessels to supply the embryo. The embryo implants completely within the uterine wall.

During this time, gastrulation is also occurring. The inner cell mass localizes to form a bilayer germ disc with an epiblast and hypoblast. The amniotic cavity forms above the epiblast, and the yolk sac forms below the hypoblast. The hypoblast layer lines the yolk sac. The extraembryonic mesoderm surrounding the embryo forms a connecting stalk, and the chorionic cavity forms around the embryo.

Between the epiblast and hypoblast layers of the germ disc, the notochord forms. The notochord is a primitive spine that ultimately becomes part of the intervertebral discs. Extraembryonic mesoderm migrates in between the epiblast and hypoblast germ layers. With this, the bilaminar germ disc becomes a trilaminar disc, with an ectoderm, mesoderm, and endoderm, and notochord in the middle.

Surface ectoderm contributes to the development of skin, hair, nails, and cornea, and gives rise to the neuroectoderm. Neuroectoderm contributes to the development of the central nervous system and retina.

Mesoderm contributes to the development of the notochord and muscle. Mesoderm and neural crest cells together are known as mesenchyme. Mesenchyme gives rise to the dermis, circulatory system, and connective tissues, including blood, bone, and cartilage.

Endoderm contributes to the development of the gut and glands. The endoderm does not contribute to any structures of the eye.

Figure Description

Early and full implantation of embryo

Key

A early implantation
B fully implanted
C uterine lining
D uterine blood vessel
E primitive yolk sac
F exocoelomic membrane
G cytotrophoblast
H hypoblast
I epiblast
J syncytiotrophoblast
K amniotic cavity
L amnion
M cytotrophoblast
N uterine lining
O uterine blood vessel
P chorionic villus
Q umbilical stalk
R chorion
S amniotic cavity
T epiderm
U mesoderm
V endoderm
W yolk sac
X chorionic cavity
Y syncytiotrophoblast

IMPLANTATION & GASTRULATION

week 2

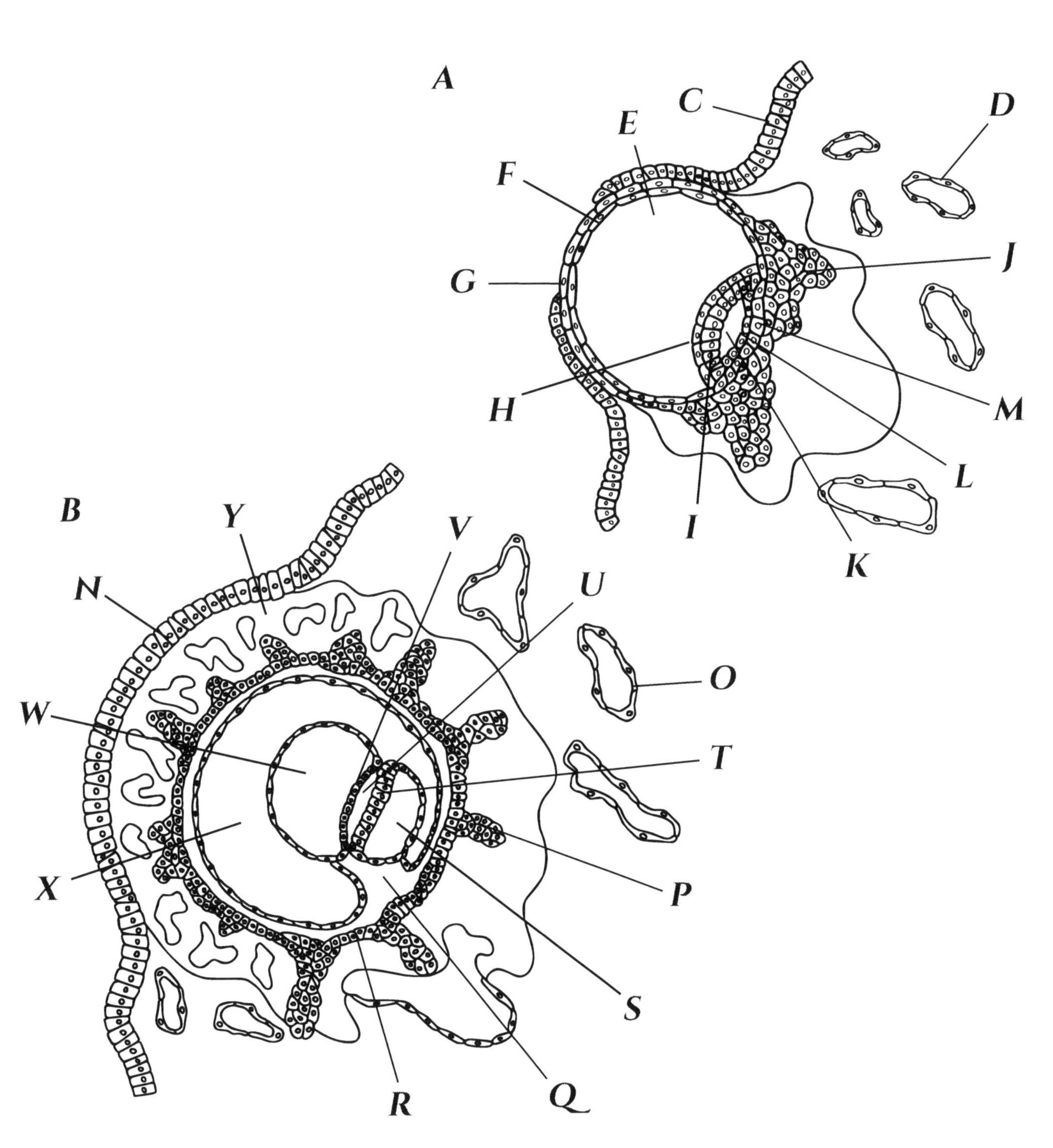

Neurulation

In week 3, the neural tube closes, in the process of neurulation. The notochord induces the ectoderm layer to thicken, forming a neural groove. The neural folds on either side of the groove move towards each other. Neural crest cells are found on the neural folds. The two neural folds meet each other and fuse in the middle region of the embryo. The neural folds then begin to fuse along the length of the embryo towards each end, leaving an opening at the cranial end, known as the anterior neuropore, and an opening at the caudal end, known as the posterior neuropore. Eventually, the entire neural tube closes. The cells of the neural tube are neuroectoderm. Neural crest cells separate from the neuroectoderm.

As the neural tube closes, mesoderm collects and forms somites on either side of the neural tube. Somites are primitive body segments.

Figure Description

Dorsal view of embryo in week 3

Key

A neural fold
B neural groove
C pericardial bulge
D amnion
E fused neural folds
F somite
G caudal neuropore

Neurulation

week 3

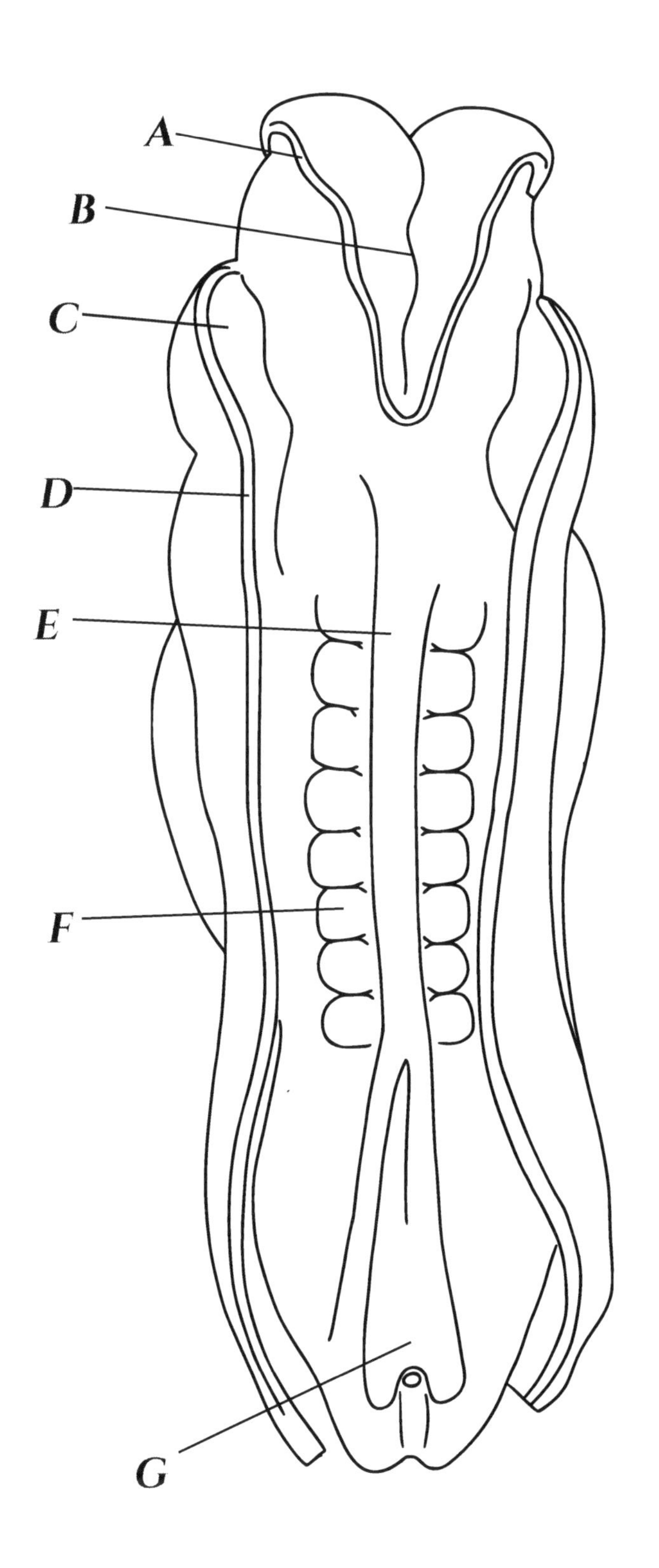

Optic Vesicle Formation

When the neural tube closes, it is in the three vesicle stage, including the prosencephalon (forebrain), mesencephalon (midbrain), and rhombencephalon (hindbrain). The neuroectoderm in the forebrain begins to grow laterally outwards, forming two optic vesicles. As the neuroectoderm reaches the surface ectoderm, it induces thickening of the surface ectoderm, known as lens placode. The optic vesicles are surrounded by mesenchyme.

Soon after, the neural tube becomes five vesicles. The forebrain differentiates into the telencephalon, which will become the cerebrum, and the diencephalon, from which the optic vesicles are connected via the optic stalk. The hindbrain differentiates into the metencephalon (pons) and myelencephalon (medulla).

Figure Description

Embryo in week 4, three vesicle stage, optic vesicle formation, adapted from Wolff, 1968.

Key

A prosencephalon, aka forebrain
B neural ectoderm
C lens placode
D mesenchyme
E optic stalk
F optic vesicle
G ectoderm
H mesencephalon, aka midbrain
I rhombencephalon, aka hindbrain
J spinal cord

Optic Vesicles

week 4

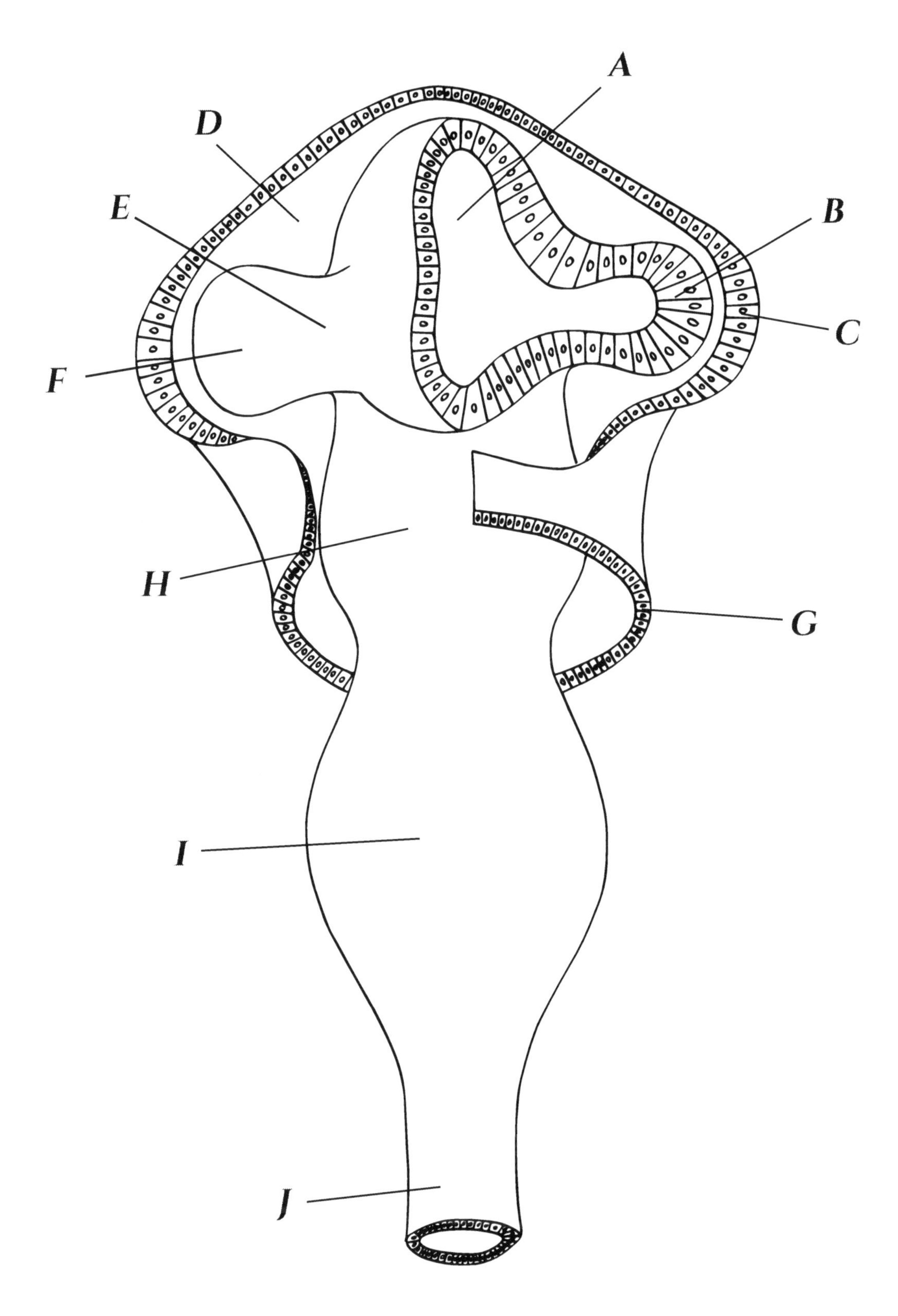

Lens Vesicle & Optic Cup

The lens placode invaginates to form a lens pit, which pinches off of the surface ectoderm to become the lens vesicle. The lens vesicle is the primitive crystalline lens, a hollow structure surrounded by ectoderm. Following separation of the lens vesicle from surface ectoderm, the surface ectoderm regenerates and forms the future corneal epithelium.

The ectoderm on the posterior surface of the lens vesicle elongates towards the anterior surface, transitioning into primary lens fibers that make up the embryonic nucleus of the crystalline lens. The ectoderm on the anterior surface becomes the anterior lens epithelium. The lens vesicle is fully filled by week 7.

The optic vesicle invaginates and becomes a two layered optic cup connected to the optic stalk. The optic stalk has an inferior (ventral) fissure, the choroidal fissure, in which the hyaloid vasculature travels to reach the developing lens. At this stage, the optic stalk is hollow, but will fill in with ganglion cell axons, becoming the optic nerve by week 8.

The outer layer of the optic cup will become the retinal pigment epithelium. The inner layer of the optic cup will become the neural retina.

Figure Description

Lateral view of early eye during embryonic week 5, adapted from Wollf, 1968.

Key

A optic stalk
B optic cup
C early retinal pigment epithelium
D early retina
E lens vesicle
F hyaloid artery
G hyaloid vein
H subretinal space
I choroidal fissure
J surface ectoderm

LENS VESICLE & OPTIC CUP

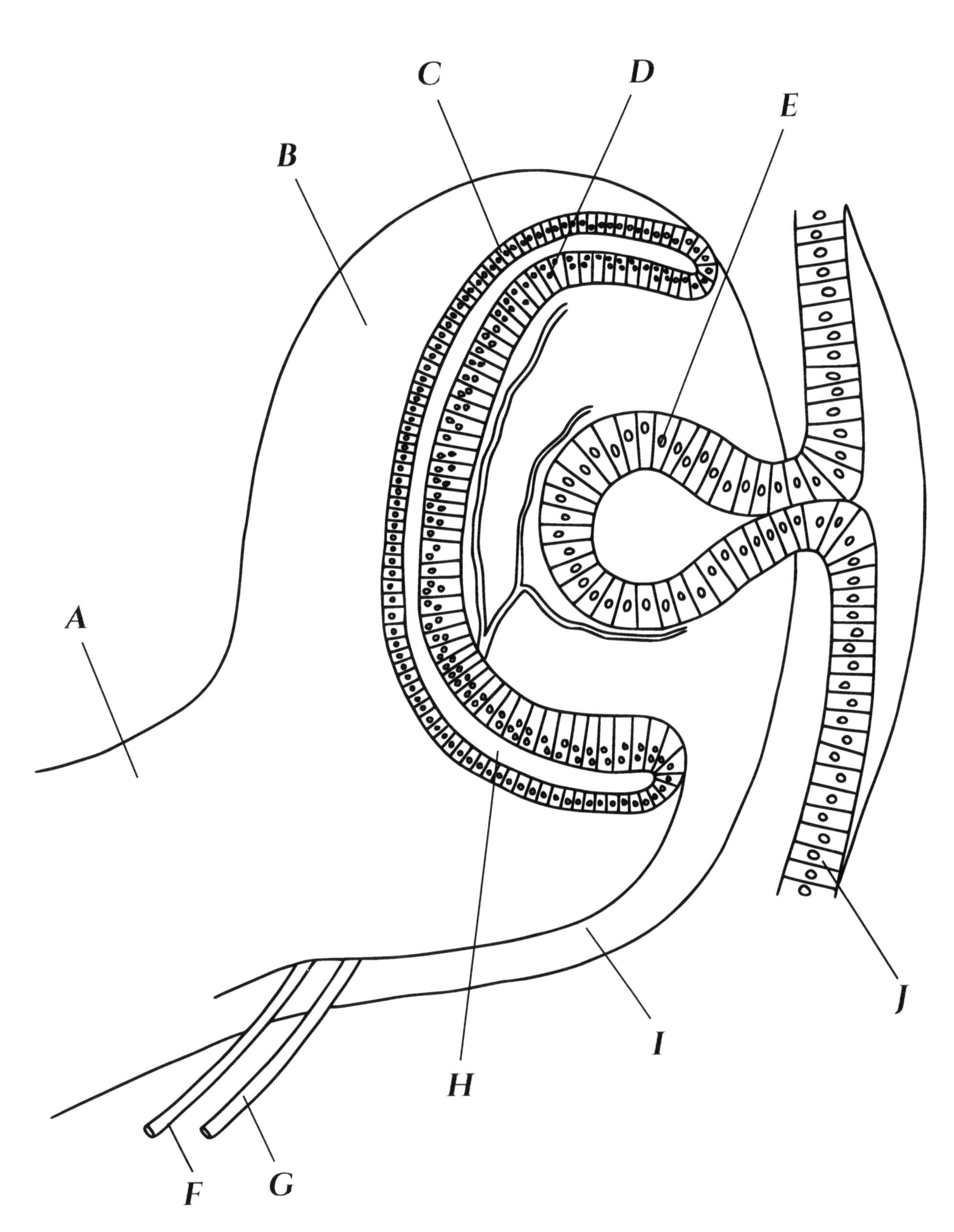

Lens and Retina Development

As soon as lens vesicle forms, it becomes vascularized by the hyaloid artery, which also vascularizes the developing retina. The hyaloid artery gains access via the choroidal fissure on ventral surface of optic stalk, extends toward the lens, giving off many small branches in vitreous chamber, called vasa hyaloidea propria. The branches surrounding the lens are the tunica vasculosa lentis. Primary vitreous is found centrally around the hyaloid artery. Secondary vitreous surrounds the primary vitreous. When the lens matures during fetal development and no longer needs a blood supply, the portion of the hyaloid artery that crosses the vitreous body degenerates, leaving Cloquet's canal, which fills in with primary vitreous.

The anterior segment forms via three waves of neural crest cells. While the corneal epithelium forms from surface ectoderm, the corneal endothelium, as well as trabecular meshwork, form from the first wave of neural crest cells. The second wave of neural crest cells form the corneal stroma and keratocytes. The third wave of neural crest cells form the ciliary body and iris.

The inner layer of the optic cup develops into the neural retina. The primitive neural retina consists of an outer neuroblastic zone and an inner acellular (or marginal) zone. Layers of cells adjacent to the subretinal space in the outer nuclear zone divide and produce waves of cells that migrate inwards toward vitreous body. As new cells reach the inner retina, they form the inner neuroblastic zone, or marginal layer of His. The two neuroblastic layers are separated by an acellular zone called the transient layer of Chievitz. Retinal cells develop in the following order: first ganglion cells, then horizontal cells, cones, amacrine cells, bipolar cells, rods, and finally, Müller cells. Photoreceptors, bipolar cells, and horizontal cells form in the outer neuroblastic zone, and ganglion cells, Müller cells, and amacrine cells form in the inner neuroblastic zone. All of the cell layers of the adult retina are apparent by the 8th month of development. However, retinal organization and foveal maturation continues over the first several years of life.

Figure Description

Lateral view of a sagittal cross section of the developing eye during embryonic week 7

Key

A optic nerve
B retinal pigment epithelium
C subretinal space
D outer neuroblastic zone of retina
E inner neuroblastic zone of retina
F central retinal artery
G hyaloid artery
H vasa hyaloidia propria
I vitreous
J eyelid bud
K early corneal epithelium (surface ectoderm)
L tunica vasculosa lentis

Lens & Retina Development

week 7

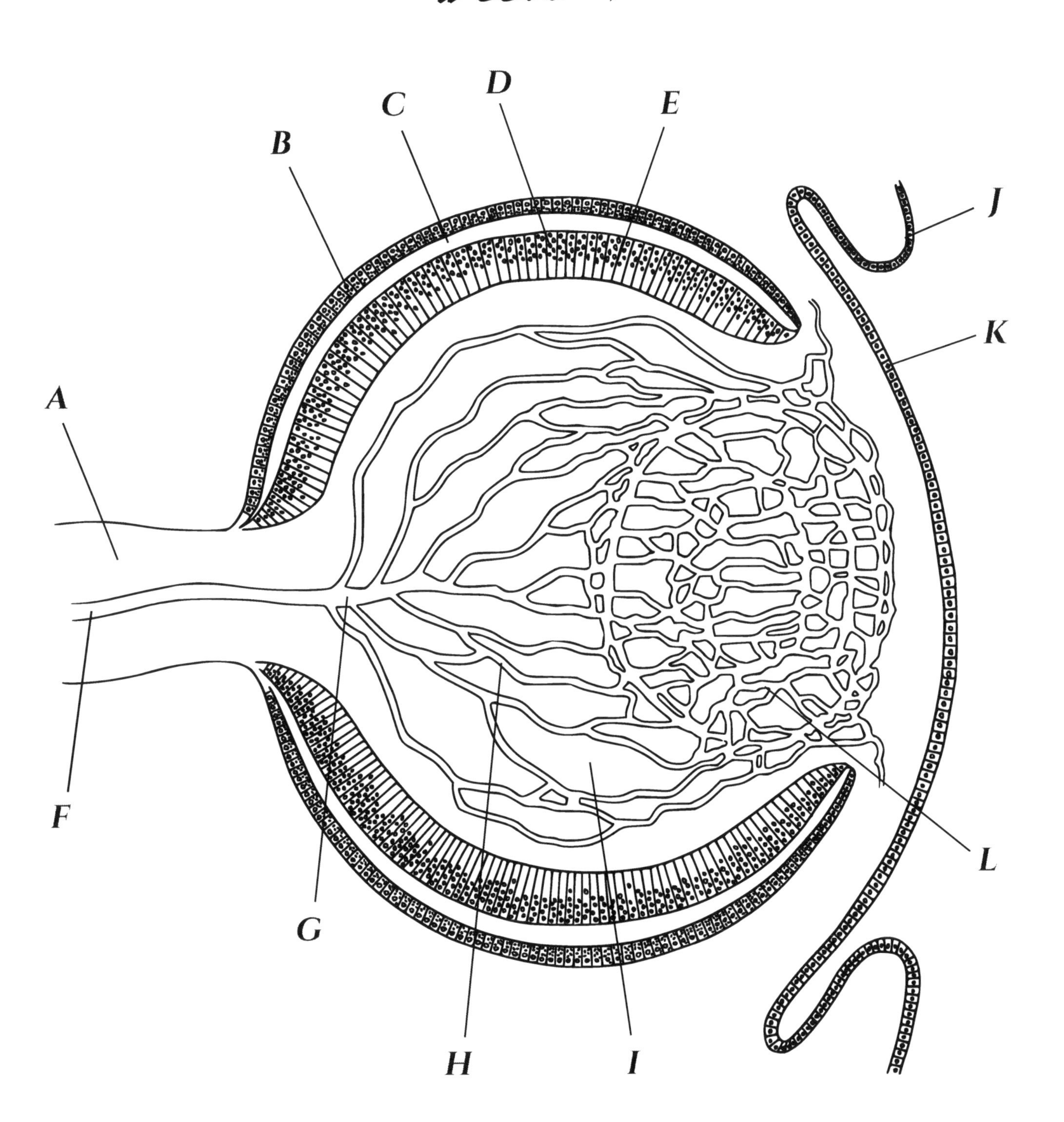

REFERENCES

Airaksinen PJ, Doro S, Veijola J. Conformal geometry of the retinal nerve fiber layer. Proc Natl Acad Sci USA. 2008;105(50):19690-5.

Anderson DR, Hoyt WF. Ultrastructure of intraorbital portion of human and monkey optic nerve. Arch Ophthalmol. 1969. 506-530.

Beier C, Zhang Z, Yurgel M, Hattar S. The projections of ipRGCs and conventional RGCs to retinorecipient brain nuclei. J Comp Neurol. 2021;529:1863-1875.

Bergmanson JPG. Clinical ocular anatomy and physiology. 27th edition, 2020.

Blumenfeld H. Neuroanatomy through clinical cases. Publisher: Sunderland, MA. Sinauer Associates, Inc. 2nd edition, 2018.

Bowd C, Weinrab RN, Williams JM, Zangwill LM. The retinal nerve fiber layer thickness in ocular hypertensive, normal, and glaucomatous eyes with optical coherence tomography. Arch Ophthalmol. 2000;118:22-6.

Cunningham DJ, Robinson A. Cunningham's text-book of anatomy. Publisher: Oxford Press Warehouse, Longon, E.C. 5th edition, 1918.

Dacey DM, Liao HW, Peterson BB, Robinson FR, Smith VC, Pokorny J, Yau KW, Gamlin PD. Melanopsin-expressing ganglion cells in primate retina signal colour and irradiance and project to the LGN. Nature. 2005;433:749–754.

Freddo TF. Chaum E. Anatomy of the eye and orbit: the clinical essentials. Publisher: Wolters Kluwer. 1st edition, 2018.

Fu Y. Liao HW, Do MTH, Yau KW. Non-image-forming ocular photoreception in vertebrates. Curr Opin Neurobiol. 2005;15(4):415-22.

Gipson IK. Goblet cells of the conjunctiva: a review of recent findings. Prog Retin Eye Res. 2016;54:49-63.

Gray H. Gray's Anatomy. Crown Publishing, 1977.

Guillery RW, Sherman SM. Thalamic relay functions and their role in corticocortical communication: generalizations from the visual system. Neuron. 2002;33 (2):163-75.

Hendry SHC, Clay RR. The koniocellular pathway in primate vision. Annu Rev Neurosci. 2000; 23:127-153.

Hogan MJ, Alvarado JA, Weddell JE. Histology of the human eye: an atlas and textbook. Published: Philadelphia, WB Saunders, 1971. ***figures adapted with permission from Elsevier*

Horner WE. Description of a small muscle at the internal commissure of the eyelids. Philadelphia J Physic Sci 1824; VIII:70-80.

Johnson SB, Coakes RL, Brubaker RF. A simple photogrammetric method of measuring anterior chamber volume. Am J Ophthalmol. 1978;85(4):469-74.

Kolb H, Fernandez E, Nelson R. Webvision: The organization of the retina and visual system. https://webvision.med.utah.edu/

Labiris G, Gkika M, Katsanos A, Fanariotis M, Alvanos E, Kozobolis V. Anterior chamber volume measurements with Visante optical coherence tomography and Pentacam: repeatability and level of agreement. Clin Exp Optom. 2009;39:772-774.

Levin LA, Nilsson SFE, Hoeve JV, Wu S, Kaufman PL, Alm A. Adler's physiology of the eye. Publisher: Saunders, 11th edition, 2011.

Liao HW, Ren X, Peterson BB, Marshak DW, Yau KW, Gamlin PD, Dacey DM. Melanopsin expressing ganglion cells on macaque and human retinas form two morphologically distinct populations. J Comp Neurol 2016;524(14):2845-2872.

Majumder PD. Anatomy of the conjunctiva. EOphtha. 2021.

Mathias RT, White TW, Gong X. Lens gap junctions in growth, differentiation, and homeostasis. Physiol Rev. 2020;90 (1):179-206.

McDougal DH, Gamlin PD. Autonomic control of the eye. Compr Physiol. 2015;5(1):439-473.

Nturibi E, Bordoni B. Anatomy, head and neck, greater petrosal nerve. StatPearls. Jan 2021.

Poirier P, Charpy A. Traité d'anatomie humaine. Paris, Masson, 1912.

Schünke Michael, Schulte E, Schumacher U, Voll MM, Wesker K. Thieme Atlas of anatomy. Head, neck, and neuroanatomy. Publisher: New York, Thieme, 2nd edition, 2016.

Skiba NP, Spencer WJ, Salinas RY, Lieu EC, Thompason JW, Arshavsky VY. Proteomic identification of unique photoreceptor disc components reveals the presence of PRCD, a protein linked to retinal degeneration. J Proteime Res. 2013;12:3010-3018.

Wolff E, Bron AJ, Tripathi RC, Tripathi BJ. Wolff's anatomy of the eye and orbit. Publisher: London; Weinheim; Tokyo: Chapman & Hall Medical, 6th edition, 1968.

INDEX

Notochord 170
Occipitofrontalis muscle 6
Oculomotor nerve, CN III 20, 136, 140, 156
Olfactory nerve, CN I 136

Made in the USA
Middletown, DE
24 October 2023